DEATH VALLEY ROCKS!

Amazing Geologic Sites in America's Hottest Park

MARLI B. MILLER
Photographs by the Author

2024
Mountain Press Publishing Company
Missoula, Montana

GEOLOGY ROCKS!
A state-by-state series that introduces readers to some of the most compelling and accessible geologic sites in each state.

© 2024 by Marli B. Miller
First Printing, July 2024
All rights reserved

Maps and figures constructed by Chelsea M. Feeney (cmcfeeney.com) from author drafts.

Cover photo: *Badlands at Zabriskie Point.*

Library of Congress Cataloging-in-Publication Data

Names: Miller, Marli Bryant, 1960- author, photographer.
Title: Death Valley rocks! : amazing geologic sites in America's hottest park / Marli B. Miller, photographs by the author.
Description: Missoula, Montana : Mountain Press Publishing Company, 2024. | Series: Geology rocks! | Includes bibliographical references and index. | Summary: "Whether you're an itinerant traveler, a seasoned geologist, or simply someone who appreciates the beauty of the natural world, Death Valley Rocks is your key to unlocking the mysteries of Death Valley National Park, one of the most enigmatic places on the planet. —Provided by publisher.
Identifiers: LCCN 2024017304 | ISBN 9780878427185 (paperback)
Subjects: LCSH: Geology—Death Valley National Park (Calif. and Nev.)—Guidebooks. | Geology—Death Valley (Calif. and Nev.)—Guidebooks. | Rocks—Death Valley National Park (Calif. and Nev.)—Guidebooks. | Rocks—Death Valley (Calif. and Nev.)—Guidebooks. | Death Valley National Park (Calif. and Nev.)—Guidebooks. | Death Valley (Calif. and Nev.)—Guidebooks.
Classification: LCC QE90.D35 M538 2024 | DDC 557.94/87—dc23/eng/20240509
LC record available at https://lccn.loc.gov/2024017304

PRINTED IN THE UNITED STATES

P.O. Box 2399 • Missoula, MT 59806 • 406-728-1900
800-234-5308 • info@mtnpress.com
www.mountain-press.com

A NOTE ABOUT SAFETY AND COLLECTING ROCKS

For all its awesome beauty, Death Valley can be a dangerous place if you're not prepared. It's rocky, steep, and bone dry; cell service is spotty at best; and medical help is a long way away. Summer temperatures frequently exceed 120°F in the shade—and winter conditions, especially when it's windy, can be penetratingly cold. At the very least, you'll want a map, sturdy shoes, a hat, and extra water (with electrolytes) when venturing away from a main road. If you're traveling a remote road, such as to the Racetrack Playa or Saline Valley, you'll need a full-sized spare tire. Do some planning beforehand and check with the park rangers about conditions. Then you'll be able to step into this incredible landscape with confidence and ease.

And please don't forget, as a national park, Death Valley belongs to everybody. Don't collect rocks, no matter how pretty they are or how much they beg you to take them home. Not only is it against the law to collect rocks within the park, it's so much better to leave them for others to enjoy.

PREFACE AND ACKNOWLEDGMENTS

Death Valley attracts people from all over the globe—even with its crazy summer heat. Its sparseness, amazing topographic relief, and wide-open vistas present a landscape found practically nowhere else. It's also one of our planet's greatest spots to study geologic processes, from erosion to faulting to Earth history. Few places have experienced as many significant geologic events as Death Valley, retained the record of those events, and display the record in such clear detail. This book includes some especially interesting and instructive sites outside the national park, which can be easily seen by visitors traveling to the park from Las Vegas.

I'm drawn to Death Valley for two reasons: its geology, obviously, and its people. Over my forty plus years of research, geotourism, and volunteering in the national park, I've made countless friends and colleagues, all of whom have supported me in one way or another. Many of them helped directly and indirectly in the writing of this book. For their endless encouragement and support, I especially thank Darrel Cowan, Birgitta Jansen, Cynthia Kienitz, Terry Pavlis, Laura Serpa, and Susan Sorrells. Darrel Cowan, Cynthia Kienitz, the Shoshone Education and Research Center (SHEAR), and Susan Sorrells all provided places to stay at one time or another during my many visits to the area.

I benefited greatly from the friends and researchers who reviewed portions of the manuscript: Ambre Chaudoin of Death Valley National Park, Darrel Cowan of the University of Washington, Judith Fierstein of the US Geological Survey, Jeffrey Goldstein of Death Valley National Park, Jeff Knott of Cal State Fullerton, Dan Larson of the University of Memphis, Chris Mattinson of Central Washington University, Tony Prave of the University of St. Andrews, Emmy Smith of Johns Hopkins University, Susan Sorrells of Shoshone, and Kevin Wilson of Death Valley National Park. Jeff Knott helped with a wide range of subjects and, above all, helped guide me through some of the more-recent findings about the Furnace Creek Basin. Mike Wells of University of Nevada, Las Vegas, and Ambre Chaudoin both took me into the field.

More than any single person, Jennifer Carey of Mountain Press turned this project into an actual book. Once again, she kindly dealt with my too many photographs and too many words (not to mention too many emails and phone calls) to create something people can enjoy and learn from—and she greatly improved the clarity of the book while preserving my voice. Our association now goes back more than fifteen years, and I'm proud to call her a friend. Chelsea Feeney took my original maps and diagrams and made them beautiful, and Jeannie Painter created the wonderful layout.

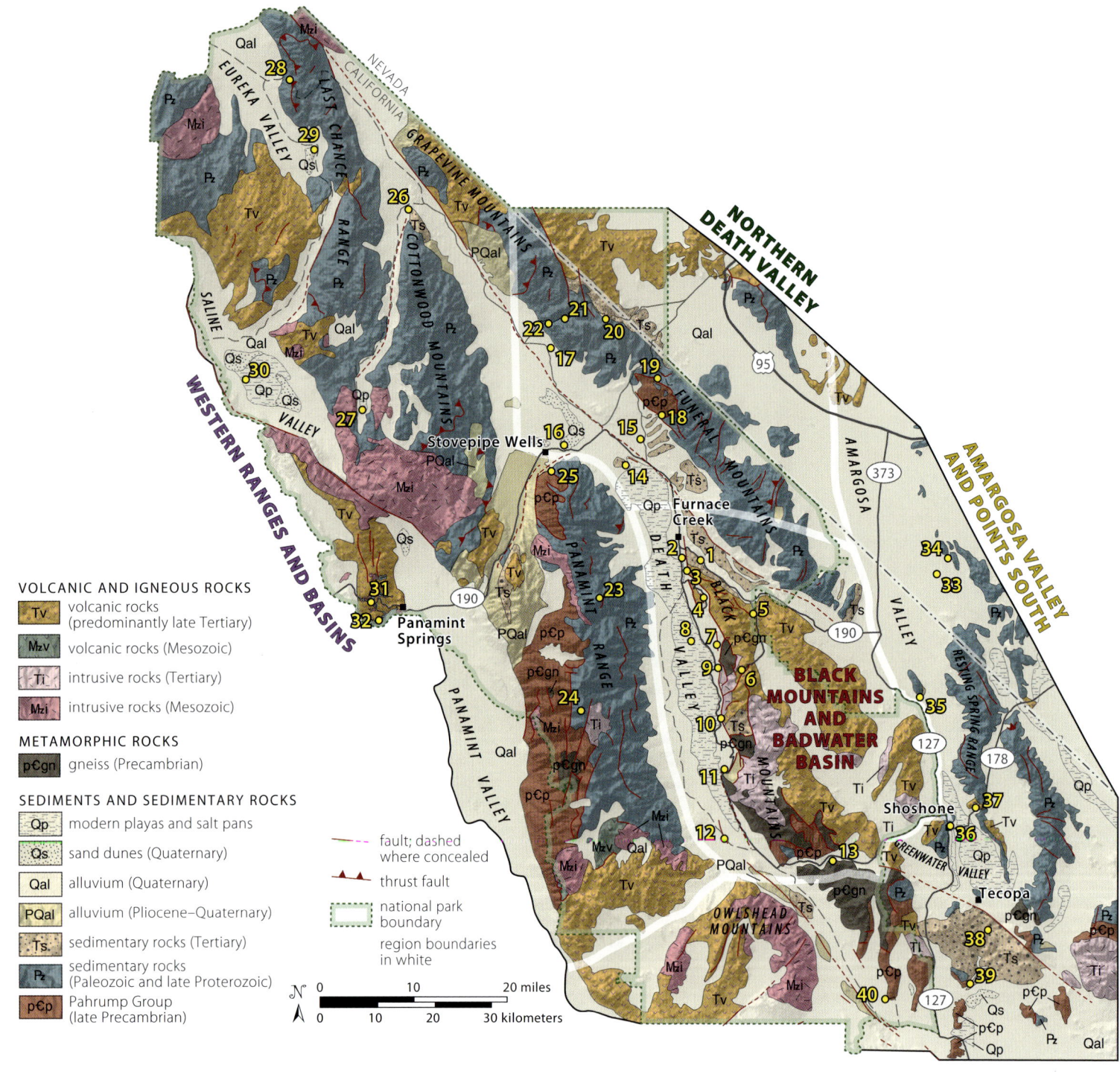

CONTENTS

A note about safety and collecting rocks   iii

PREFACE AND ACKNOWLEDGMENTS   iii

A BRIEF GEOLOGIC HISTORY OF DEATH VALLEY   1
    Fault Types   2
    Precambrian Era   6
    Paleozoic Era   9
    Mesozoic Era   11
    Cenozoic Era   11

BLACK MOUNTAINS AND BADWATER BASIN   15
    1. Zabriskie Point   16
    2. Golden Canyon   18
    3. Gower Gulch   20
    4. Artists Drive   22
    5. View of Ryan   26
    6. Dantes View   27
    7. Natural Bridge Canyon   30
    8. Devils Golf Course   34
    9. Badwater Basin   35
    10. Black Mountains Fault Zone   38
    11. Mormon Point   39
    12. Cinder Hill   42
    13. Exclamation Point   44

NORTHERN DEATH VALLEY   47
    14. Salt Creek Hills   48
    15. Beatty Bar   50
    16. Mesquite Flat   52
    17. Debris Flow Fan   54
    18. Keane Wonder Mine   56
    19. Monarch Canyon   58
    20. Red Pass   62
    21. Titus Canyon Anticline   64
    22. Titus Canyon Jigsaw Puzzle   66

WESTERN RANGES AND BASINS   69
    23. Aguereberry Point   70
    24. Telescope Peak   72
    25. Mosaic Canyon   75
    26. Ubehebe Crater Field   78
    27. Racetrack Playa   81
    28. Hanging Rock Canyon   84
    29. Eureka Dunes   86
    30. Saline Valley   88
    31. Father Crowley Vista Point   92
    32. Darwin Falls   96

AMARGOSA VALLEY AND POINTS SOUTH   99
    33. Ash Meadows National Wildlife Refuge   100
    34. Devils Hole   102
    35. Eagle Mountain   106
    36. Shoshone   109
    37. Roadcut at Resting Spring Pass   112
    38. China Ranch   115
    39. Mouth of the Amargosa River Canyon   118
    40. Saratoga Spring   121

GLOSSARY   124

SOURCES   127

INDEX   133

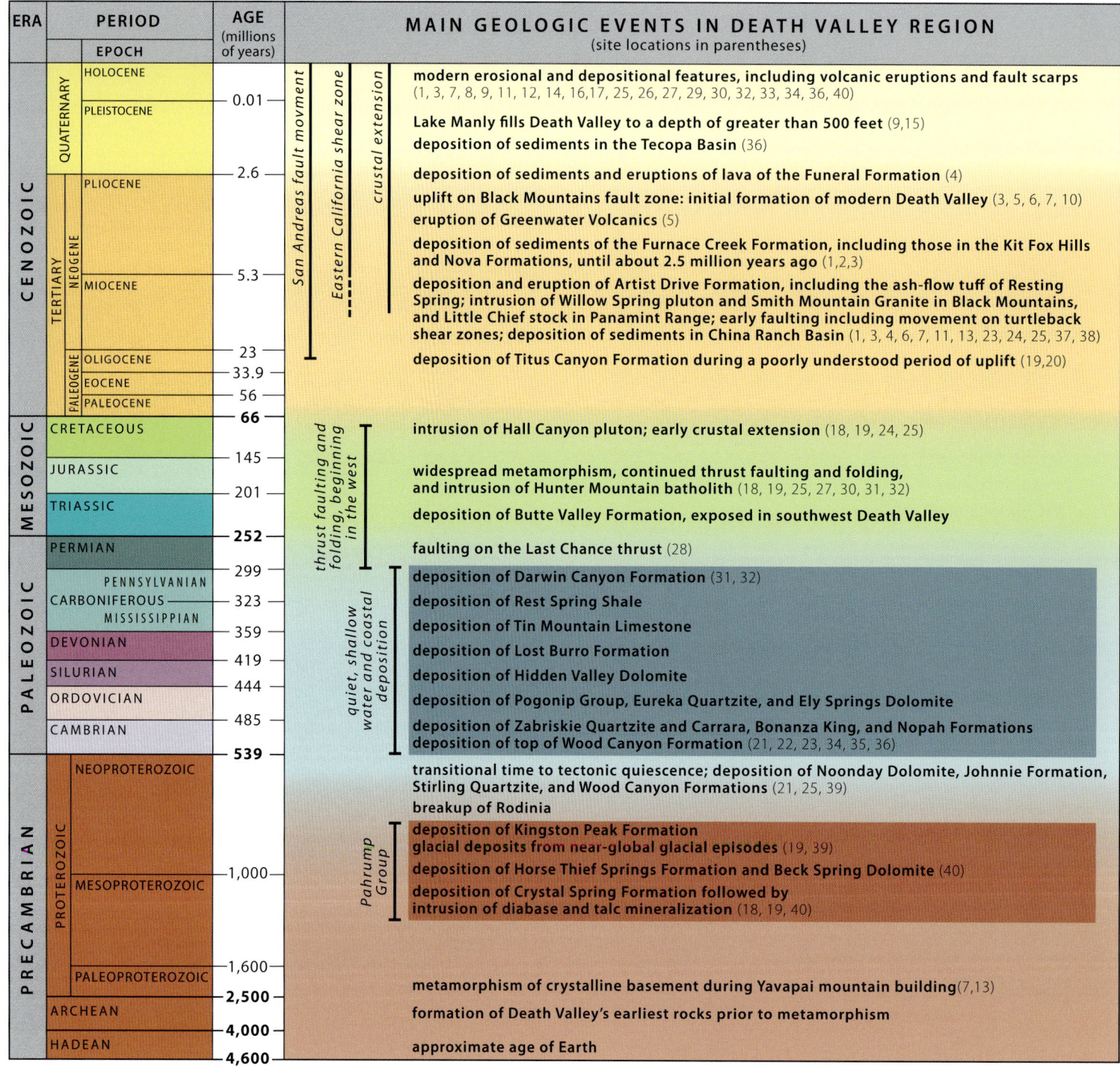

A BRIEF GEOLOGIC HISTORY OF DEATH VALLEY

You don't have to travel far in Death Valley to witness most of the incredible events that shaped—and continue to shape—western North America. From Precambrian glaciation to modern-day mountain building, these events left their signatures in Death Valley's rock record and landscape. And because of Death Valley's extreme aridity, many of those signatures are in plain sight. Indeed, the region has long proved a wonderful destination for countless geology field trips, geology-seeking tourists, and researchers.

Called *Tüpippüh* by the Timbisha Shoshone who have lived in the region for over a thousand years, Death Valley offers desert in its most extreme. Summer temperatures regularly exceed 120°F, and typical annual rainfall reaches barely more than 2 inches per year. These dry, hot conditions result from the region's low elevation as well as its position in the rain shadow of the Sierra Nevada. As moisture-laden air blows off the Pacific Ocean and rises over the Sierra Nevada, it cools and loses its water. By the time it reaches the east side of the mountains, the air is cool and dry.

With an area of more than 5,200 square miles in southeastern California and southwestern Nevada, Death Valley National Park is the largest national park in the contiguous United States. Its mountains and valleys showcase some of the greatest relief in the Lower Forty-Eight. Badwater, at 282 below sea level, lies only 12 miles away from Telescope Peak, which reaches 11,049 feet above sea level. Death Valley is an unusually dramatic example of the Basin and Range Province, a region of near-parallel mountain ranges and intervening valleys that stretches from California's Sierra Nevada to Utah's Wasatch Mountains.

Movement along faults uplifted the mountains and dropped the valleys. These fractures in the Earth's crust, called normal faults, formed because the crust in the Death Valley area—as well as the Basin and Range in general—is being pulled, or stretched, apart by tectonic forces. Death Valley lies along the margin of the North American tectonic plate. As the continental plate moves southwest, the Pacific plate moves north. The boundary between the two plates is the San Andreas fault, which has right-lateral, strike-slip movement. This plate-bounding fault is one of many similar faults in eastern California and western Nevada that stretch, or extend, the region.

In the Death Valley area, the crust is pulled apart between different overlapping strands of right-lateral, strike-slip faults. For

Right-lateral, strike-slip motion on the San Andreas fault contributes to the stretching apart of the Basin and Range Province and the dropping of deep basins such as Death Valley. The Cascadia subduction zone lies north of the San Andreas fault.

FAULT TYPES

Death Valley offers all types of faults. The main faults active today are normal faults and strike-slip faults, but at different times in the past, reverse and thrust faults dominated the region. Normal faults, which mark the west side of the Black Mountains, allow the crust to extend. Younger rock moves downward along the upper surface of the fault so it's emplaced on top of older, deeper rock. Some normal faults are inclined at unusually low angles and known as detachment faults. Higher-angle normal faults above a detachment typically end where they run into the detachment and transfer their slip onto the detachment surface. Many of the larger normal faults in Death Valley, such as the Black Mountain turtleback faults and the Boundary Canyon fault, are detachment faults.

In contrast to normal faults, thrust and reverse faults form where the crust contracts. They typically place deeper, older rock on top of younger rock. Because they form by compression, thrust faults are typically associated with folded rocks.

Instead of moving rock up and down like normal and thrust faults, strike-slip faults displace rock laterally. Death Valley's strike-slip faults are mostly right-lateral faults, in which the blocks on opposite sides of the fault slide to the right relative to each other. In left-lateral faults, the opposing blocks move to the left.

A. A normal fault moves rock above the fault down, placing it on top of older rock. Arrows indicate the direction of movement.

B. In this strike-slip fault, the band of rocks at the middle right has moved to the right relative to the band below. Arrows indicate the direction of movement.

C. In this thrust fault, formed during compression, older rock moved up over younger rock. Arrows indicate the direction of movement.

D. The Badwater Turtleback fault, shown here in a locality outside of Natural Bridge Canyon (site 7), is a detachment fault. Notice how the two high-angle normal faults terminate at the detachment. Both the detachment and the high-angle faults may slip at the same time.

example, central Death Valley is moving northwestward along the Northern Death Valley fault zone and southeastward along the Southern Death Valley fault zone, causing the valley to stretch apart. These two faults, along with Hunter Mountain–Saline Valley fault, the Owens Valley fault, and several others, constitute the Eastern California shear zone. Like the San Andreas fault to the west, these faults are products of right-lateral shear along the North American–Pacific plate margin. Collectively, faults of the Eastern California shear zone account for about a quarter of the relative motion along the plate margin.

We can see evidence for this ongoing crustal extension and mountain building in Death Valley's landscape, particularly along the east side of the valley floor. There, the Black Mountains rise abruptly and steeply more than 1 mile in elevation, with no intervening foothills. A normal fault separates the mountains from the valley floor. In some places, such as near Badwater and just north of Golden Canyon, normal faults break through the recently deposited gravel to produce fault scarps. These small steps in the land surface displace young gravel deposits, indicating recent movement. Along the range front, you can also see triangular

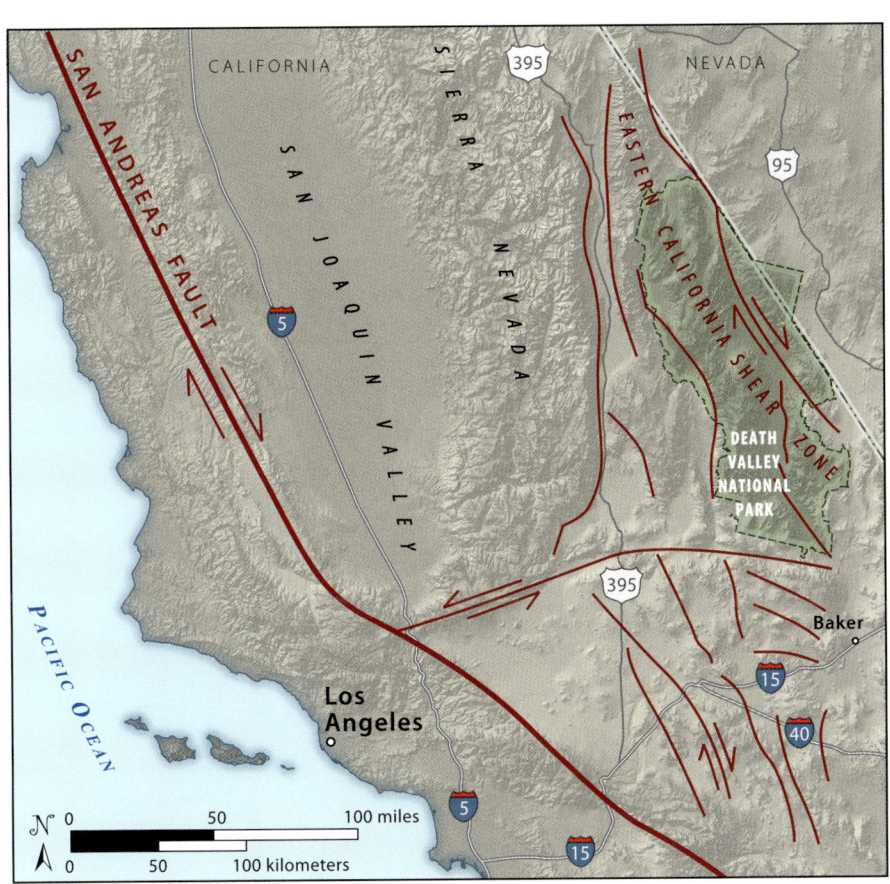
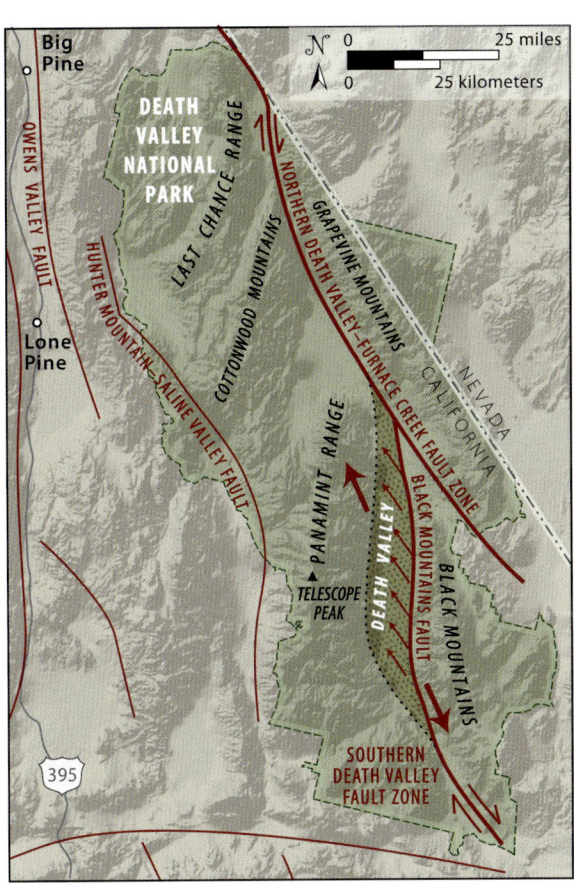

Map showing some of the principal faults of the Eastern California shear zone. Death Valley National Park is shaded green. Enlarged version on the right shows how right-lateral slip on the Northern Death Valley–Furnace Creek fault and the Southern Death Valley fault pulls apart the area between them. The Black Mountains fault is a normal fault, dropping Death Valley downward as it is pulled apart.

facets, which form because ridges that would otherwise descend gently to the valley floor are cut off by the fault zone.

When viewed face-on from the west, the shapes of some of the canyons exiting the Black Mountains resemble wineglasses, indicating recent faulting. Alluvial fans form the bases of the wineglasses, the narrow and steep canyon mouths form their stems, and the wider, gentler, upper parts of the canyons form their bowls. The active faulting along the mountain front continually uplifts the canyon mouths to keep them steep and narrow. Behind the fault, the canyons erode gradually into the wider, gentler bowls. Storm runoff carries the eroded debris through the canyons and deposits the sand, gravel, and boulders in the alluvial fans at the edge of the valley floor. On the east side of the Black Mountains, however, there is little active faulting and the rocks tilt gently eastward into the Greenwater Valley.

The Panamint Range on the west side of Death Valley, although much higher in elevation than the Black Mountains, displays roughly the same shape. On its west side, the range rises steeply along a normal fault zone that lines the edge of Panamint Valley to its west. On the Death Valley side, however, the Panamint Range descends much more gradually toward the east.

The asymmetries of the Black Mountains and Panamint Range—precipitously steep on their faulted western sides and comparatively gentle on their unfaulted eastern sides—are typical of tilted fault blocks. A normal fault on one side uplifts these mountain-scale blocks of crust on one side, tilting them away from the fault. Similar tilted fault blocks typify most places that experience crustal extension and are found throughout the Basin and Range Province. Death Valley, with its position along the western edge of the Basin and Range, displays some of the most-dramatic tilted fault blocks in the world.

The extreme desert landscape of the valley floor displays an amazing variety of landforms in Death Valley. Fierce winds pick up and carry fine-grained sand and silt from the alluvial fan surfaces and deposit it as sand dunes. The windblown sand abrades rocks that sit in places unprotected from the wind, creating what are known as ventifacts. Salt covers much of the valley floor, some recently precipitated from evaporating water carried there during storm events and more remaining from when Lake Manly evaporated. The lake filled Death Valley to a depth of greater than 500 feet during the ice ages of the Pleistocene Epoch. You can see shorelines of this lake perched along the front of the Black Mountains, and a former spit is now a prominent gravel deposit along Beatty Cutoff Road some 13 miles north of Furnace Creek. Death Valley even hosts some young volcanic landforms: the Ubehebe Crater field in the northern part of the park and a cinder cone in the southern part.

Wineglass canyons and triangular facets along the front of the Black Mountains, south of Badwater. The dashed line outlines one of the triangular facets.

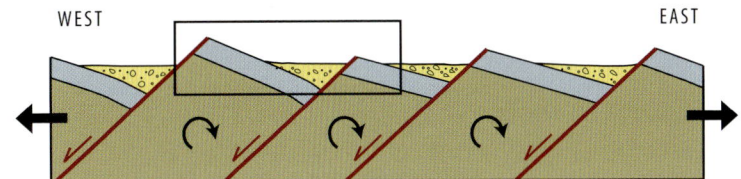

Tilted fault blocks in the Death Valley area rise directly behind a normal fault on their west sides and tilt to the east. The rectangle highlights a general representation of the higher Panamint Range on the west and the lower Black Mountains on the east.

Parallel ranges and intervening basins as seen looking east from Dantes View. The Spring Mountains in Nevada form the high range on the skyline.

Ventifacts, rocks sculpted by windblown sand, sit on an exposed surface. The steep, western front of the Black Mountains rises on the left skyline.

With only sparse vegetation and practically no soil development to soften the landscape, dramatic colors and unusual erosional features abound in Death Valley. Most of the red, oranges, yellows, and greens in the rocks derive from hydrothermal fluids that accompanied periods of volcanic activity and mountain building during Death Valley's eventful geologic history.

PRECAMBRIAN ERA
(4.6 billion to 539 million years ago)

Death Valley's oldest rocks are part of the basement, the deepest rock level that forms the region's crustal foundation. These metamorphic rocks consist of gneiss and schist and show up in the central and southern Black Mountains and in the southern Panamint Range. Prior to metamorphism, about 1.7 billion years ago, they originated as both sedimentary and igneous rock, parts of which may date to more than 2.5 billion years ago, the boundary between the Archean and the younger Proterozoic Eons. Of the sites described in this book, you see basement rock exposed in Natural Bridge Canyon (site 7) and Exclamation Point (site 13).

Death Valley's basement resembles other basement rocks exposed in scattered localities throughout the Mojave Desert as far east as the California–Arizona border. Together, these rocks form a body of crust called the Mojave terrane, which was accreted to the core of North America and metamorphosed about 1.7 billion years ago during the Yavapai mountain building event. The oldest sedimentary rocks of Death Valley belong to the Pahrump Group, which was deposited on top of the basement about 500 million years later, between about 1.2 billion and 600 million years ago.

The latter part of this 500-million-year period was undoubtedly marked by uplift and erosion because the basement rocks had to be exposed at Earth's surface by the time the Pahrump Group sediments were deposited. From nearby regions, we can infer that it coincided with Grenville mountain building at about 1.2 billion years. This mountain building event occurred because a continental collision stretching from west Texas to New England added basement rock to the growing North American continent. These plate tectonic collisions marked early stages in the assembly of the supercontinent Rodinia.

The Pahrump Group was deposited over a lengthy time period, extending from the final assembly of Rodinia to its

Basement gneiss in the Black Mountains. Hammer head for scale.

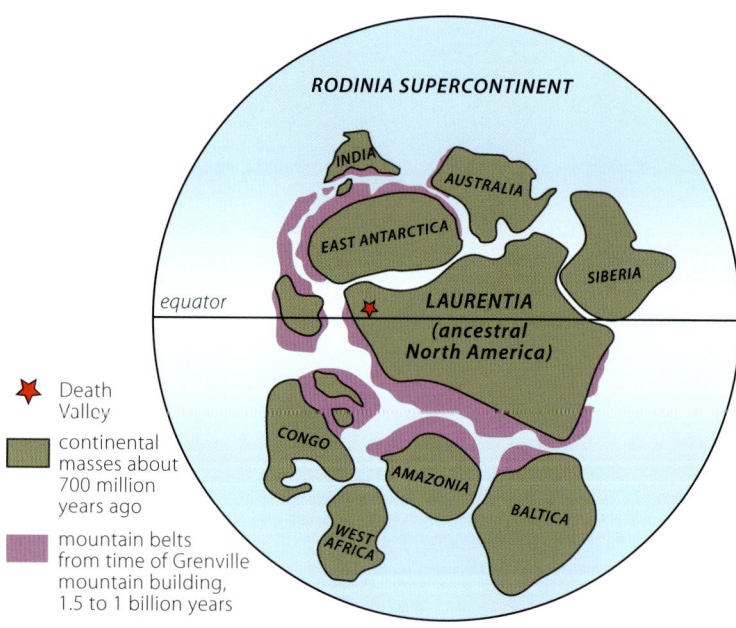

The supercontinent Rodinia was a collection of continental fragments that came together before deposition of the Pahrump Group, likely during Grenville mountain building. —Modified from Hoffman, 1991

subsequent breakup. The Pahrump is beautifully exposed in the southern Black Mountains near Saratoga Spring (site 40), in the central Black Mountains as fault slices in the Amargosa Chaos of Exclamation Point (site 13), and on the western slopes of the Panamint Range. In the northern Funeral Mountains near the Keane Wonder Mine (site 18) and Monarch Canyon (site 19), you can see highly metamorphosed rock of the Pahrump Group.

The Pahrump Group consists of four parts, from oldest to youngest: the Crystal Spring Formation, the Horse Thief Springs Formation (formerly the upper part of the Crystal Spring), the Beck Spring Dolomite, and the Kingston Peak Formation. The Crystal Spring Formation consists mostly of river-deposited sandstone, siltstone, and shale that grade upward into shallow-ocean limestone and dolomite. These sedimentary rocks are intruded by diabase, an igneous rock with roughly the same composition as the extrusive rock basalt that intruded into shallow levels of the crust but did not reach the surface. In Death Valley, most of the diabase forms sills, which solidify after magma intrudes parallel to the existing rock layering. Chemical reactions between the sills and dolomite produced talc deposits, the mining of which played an oversized role in Death Valley's human history.

After a hiatus of about 200 million years, the overlying Horse Thief Springs Formation and Beck Spring Dolomite were deposited over the top of the Crystal Spring Formation. Deposited in coastal and warm shallow-ocean environments, they mark a more stable period of Rodinia. Above them, the widely variable Kingston Peak Formation marks the onset of Rodinia's breakup due to rifting. As the supercontinent was stretched apart in the rifting process, the crust thinned and a basin formed that eventually filled with seawater. The Kingston Peak Formation also contains thick sequences of diamictites, fine-grained rocks with numerous suspended cobbles and boulders. Some of these diamictites appear to have a glacial origin—meaning they formed when icebergs dropped entrained rocks to the seafloor or when till deposits sloughed into

View of the west side of the Funeral Mountains, which are composed mainly of the Pahrump Group. The arrow points to the approximate location of the trailhead for the Keane Wonder Mine (site 18).

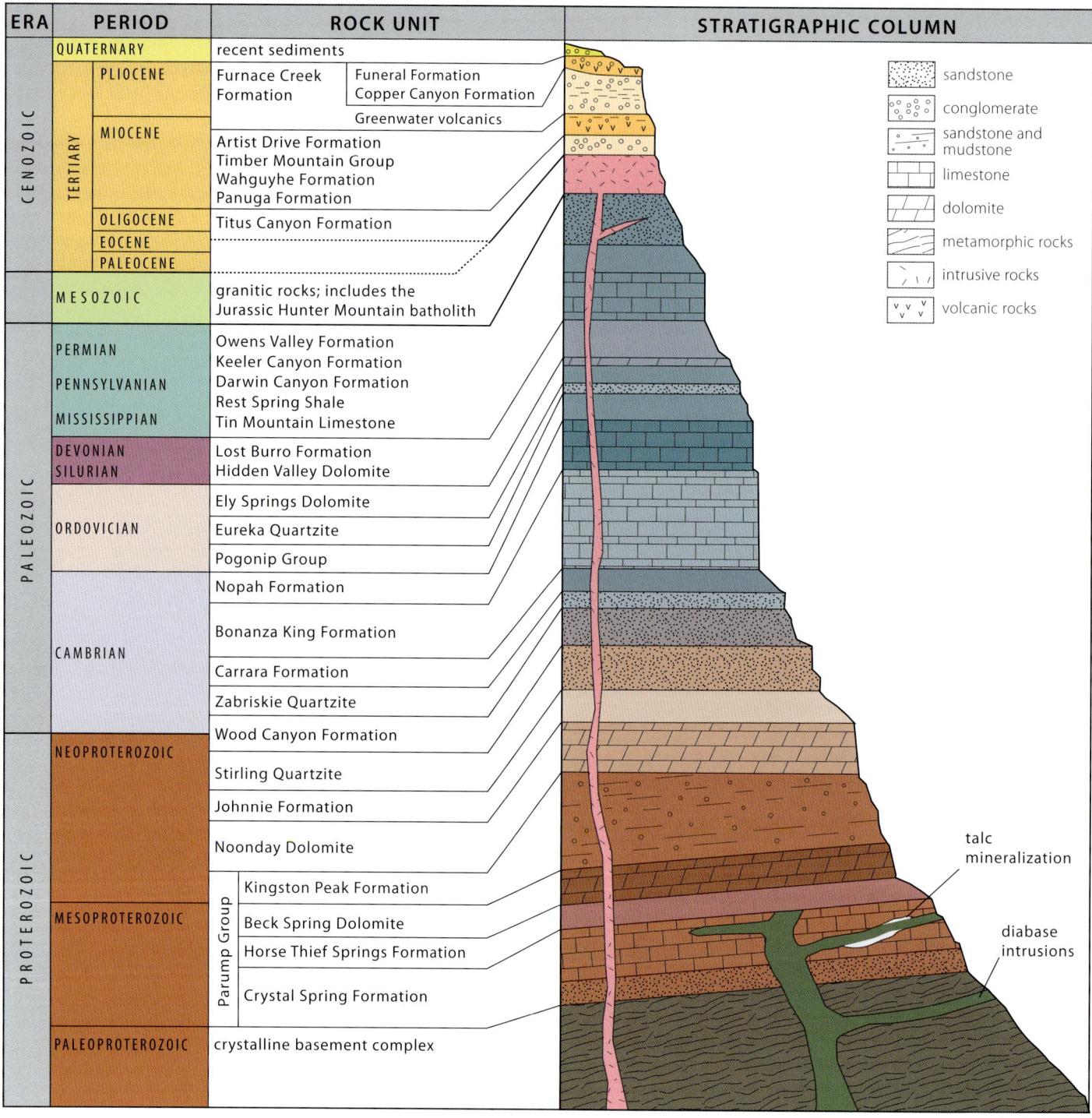

Stratigraphic section showing the main rock units of the Death Valley area.

This talc mine and white talc mineralization in southern Death Valley is situated at the contact of greenish-black diabase (lower) and brownish dolomite of the Crystal Spring Formation (upper). Note the sand dune in the foreground.

deep seawater—and provide important evidence to support the hypothesis of worldwide glaciation between 720 and 635 million years ago (visible at site 39).

A remarkable sequence of four Late Proterozoic rock units lies directly above the Pahrump Group. Deposited in environments that ranged from shallow oceans to rivers and deltas, this sequence starts with the Noonday Dolomite at the bottom, followed by the Johnnie Formation and the Stirling Quartzite, and ends with the lower third of the Wood Canyon Formation. These rocks reflect a transitional time from the continental rifting recorded by the Kingston Peak Formation to the tectonically quiet period marked by Paleozoic rocks. These rocks are highly faulted in the Amargosa Chaos area of the southern Black Mountains, and they have experienced various degrees of metamorphism in the northern Funeral Mountains and Panamint Range.

The Noonday Dolomite appears to have formed in response to the most-recent worldwide glacial episode, as recorded in the underlying Kingston Peak Formation. Rocks of similar age exist the world over and figure prominently in studies of the glacial period's aftereffects. The Johnnie, Stirling, and Wood Canyon Formations each display a great deal of variability as well as coarse-grained detritus, suggesting they were deposited during times of faulting and the final stages of the rifting of Rodinia. Moreover, the Wood Canyon Formation was deposited through time across the Precambrian–Cambrian boundary, so it contains fossils of strange soft-bodied sea creatures before and during the explosion of life at the beginning of the Cambrian Era.

PALEOZOIC ERA
(539 to 252 million years ago)

Beginning with the upper two-thirds of the Wood Canyon Formation, the Paleozoic rock sequence of the Death Valley region is unusually thick and complete. Its well-preserved units accumulated on a gradually subsiding continental margin. The sequence consists of about a dozen different rock units, which are mostly limestone, dolomite, or shale of varying shades of gray as well as a few distinctive sandstone units. Their bedding gives a striped appearance to all of Death Valley's mountain ranges with the exception of the Black Mountains, which are practically devoid of Paleozoic rocks.

Both limestone and dolomite form through the precipitation of carbonate, largely brought on by biological activity such as the growth of algae or the production of shells. For carbonate to be precipitated, the water must be shallow enough for light to reach the seafloor to drive photosynthesis. The presence of limestone and dolomite allows geologists to infer that the Paleozoic ocean was shallow. In many places, these rocks also preserve abundant fossils of invertebrates, such as trilobites, brachiopods, and crinoids.

Several of the older Paleozoic rock units deserve special mention. The Cambrian-age Bonanza King Formation, which reaches some 2,500 feet in thickness, consists largely of alternating beds of gray and dark-gray dolomite and minor limestone. It dominates the scenery of the Funeral Mountains near the national park entrance on CA 190, the lower 4 miles of Titus Canyon, the front of the Grapevine Mountains, and the hills near Devils Hole, as well as much of the Panamint Range and Cottonwood Mountains.

The Cambrian Zabriskie Quartzite and Ordovician Eureka Quartzite are also prominent units, appearing as distinctive white or pink stripes on the mountainsides. Neither rock unit is typically metamorphic, as implied by the quartzite name, but instead consists mostly of strongly cemented, nearly pure quartz sandstone, making it hard and very resistant to erosion. These sandstone units represent times of sea level lowering when sand was more likely to be deposited than carbonate. Parts of the Zabriskie Quartzite even show evidence for being deposited on land.

The Bonanza King Formation, pictured near Devils Hole (site 34), consists of a thin-bedded unit beneath a thick-bedded unit. It was deposited in a shallow sea in Cambrian time.

MESOZOIC ERA
(252 to 66 million years ago)

Beginning toward the end of the Paleozoic Era and continuing through most of the Mesozoic Era, the Death Valley area experienced crustal compression brought on by a change in plate motions. Whereas the stable continental shelf drifted away from the nearest plate margin during most of the Paleozoic, it actively converged with another plate in the Mesozoic. While details of the transition are still debated, researchers agree the convergence caused mountains to rise along thrust faults, some sedimentary rocks to become buried deeply enough that heat and pressure metamorphosed them, and a variety of granite-like magmas to intrude the existing Paleozoic and Proterozoic rocks.

Thrust faults and related folds affect the rocks in the ranges surrounding Death Valley. The oldest thrusts are those in the western parts of the park, including the Permian-age Last Chance thrust in Hanging Rock Canyon (site 28), and the younger, Jurassic-age Lemoigne thrust visible from Father Crowley Vista Point (site 31). By contrast, the thrusts in the Grapevine and Funeral Mountains were likely active during the rest of the Jurassic and Cretaceous Periods of the Mesozoic Era.

The thrust faulting buried Proterozoic and Paleozoic sedimentary rocks to depths as great as 6 miles in the Panamint Range and 15 miles in the Funeral Mountains. In both places, the high temperatures associated with the burial transformed the original shales into schists, the sandstones into quartzites, and the limestones and dolomites into marbles. Both the Keane Wonder Mine (site 18) and Monarch Canyon (site 19) in the Funeral Mountains are outstanding places to see these beautiful metamorphic rocks up close.

Plate subduction along North America's west coast produced numerous bodies of Jurassic- and Cretaceous-age granites that intrude the Paleozoic rock on the west side of the park. The largest, the Jurassic-age Hunter Mountain batholith, covers an area of 100 square miles, mostly southeast of Racetrack Playa (site 27). The much-smaller Skidoo and Hall Canyon plutons intruded during the Cretaceous Period. Mineral-laden fluids derived from the crystallizing Skidoo pluton invaded the local country rock to precipitate gold-rich quartz veins, which were mined at the Skidoo site from 1906 to 1917. Other Mesozoic-age granites include the Darwin quartz diorite exposed at Darwin Falls (site 32) and those in the Owlshead Mountains near the south end of the national park.

Granitic rocks of the Hunter Mountain batholith near Racetrack Playa (site 27).

CENOZOIC ERA
(66 million years ago to present)

Geologists know very little about the first 50 million years of the Cenozoic in Death Valley, at least partly because much of that record was eroded away during the last 15 million years or so of extensional mountain building. However, the Titus Canyon Formation, deposited from about 37 to 30 million years ago, contains abundant river sediments and landslide debris, suggesting a period of uplift. Red Pass (site 20) explores features of crustal extension seen in the Titus Canyon Formation. The lower part of the Titus Canyon Formation yielded a titanothere fossil, as well as fossils of horses and rhinoceroses.

Modern crustal extension in Death Valley has its roots some 16 to 14 million years ago with the initiation of normal faulting that uplifted the Black Mountains (sites 7 and 10). Starting about 11 million years ago, magmas of the Willow Spring pluton, the Smith Mountain Granite, and some unnamed granites moved upward

Gently east-tilted red- and tan-colored volcanic rocks of the Artist Drive Formation near the crest of the Black Mountains looking north from near Dantes View. The black rock in the middle ground is 4-million-year-old basalt of the Greenwater Range. Beyond that, the striped rocks in the background are Paleozoic rock of the Funeral Mountains. The Furnace Creek fault separates the Paleozoic rock from the basalt.

into the extending crust and intruded the basement rock of the Black Mountains. These rocks now form much of the range from Smith Mountain to near Dantes View (site 6).

Perhaps most striking, however, are the multicolored volcanic rocks that blanket much of the top of Black Mountains. Most of these deposits began erupting soon after the intrusions and continued until about 6 million years ago. The Shoshone Volcanics, which cover most of the area south of Dantes View, and the Artist Drive Formation, north of Dantes View, consist of rhyolitic lavas and explosive deposits. The Artist Drive Formation also contains abundant volcanic-rich sedimentary rocks and a small fraction of dark-gray basaltic lava flows. Their beautiful colors derive from the original composition of the rock—most notably its iron content—and from chemical alteration after the rock formed.

The Furnace Creek Formation rests on top of the Artist Drive Formation. Its 7,000 feet of volcanic-rich sedimentary rock accumulated from about 6 to 2.5 million years ago. The formation includes some basaltic lava flows and hosts most of Death Valley's borate deposits. Spectacular exposures of conglomerate of the Furnace Creek Formation line either side of CA 190 between the Furnace Creek Inn and Zabriskie Point. The beautiful sea-green color is the product of alteration of volcanic ash to the clay-like mineral celadonite. At Zabriskie Point, Golden Canyon, and Gower Gulch (sites 1, 2, and 3), you can see badlands eroded into mostly fine-grained sandstone and siltstone.

The Furnace Creek Formation was deposited in the Furnace Creek Basin, which developed along the southwestern edge of the Furnace Creek fault zone. The basin extended southeastward at least as far as today's Eagle Mountain (site 35) in the Amargosa Valley. From the earliest deposition of the Artist Drive Formation to about 3.5 million years ago, the basin was a throughgoing river and temporary lake system that received sediments from as far away as the Darwin Plateau, west of Panamint Valley.

Modern Death Valley began to form about 3.5 million years ago as the Black Mountains began rising and disrupted the drainage into the Furnace Creek Basin. The younger parts of the Furnace Creek Formation and the overlying Funeral Formation—both of which contain material shed from the mountains onto alluvial fans—record this transition. A particularly striking exposure behind the old mining camp of Ryan (site 5) shows flat-lying basalt of the Greenwater Range overlying faulted and tilted Artist Drive and Furnace Creek Formations. The basalt, about 4 million years old, is gently folded overall, but its comparative lack of deformation indicates a waning of tectonic activity in this part of the Furnace Creek Basin by the time it was erupted. In lower Furnace Creek Wash, you can see tilted Furnace Creek Formation overlain by flat-lying Funeral Formation.

The climate cooled during much of the Pleistocene Epoch, beginning about 2.6 million years ago, and, occasionally, lakes filled low-lying areas throughout the region. In Death Valley, the largest was Lake Manly, named after William Manly, one of the original 49ers, the pioneers who traveled to California in 1849 for the Gold Rush. Lake Manly was at its highest sometime between 180,000 and 128,000 years ago. During this time, it possibly connected to other lakes as far away as the Owens Valley. Pupfish, a group of tiny ray-finned fish, interacted with each other across the entire region. As the climate dried and the connections between their habitats were lost, they evolved into separate species that now live in small, isolated locations. A smaller lake occupied much of Death Valley from 35,000 to 10,000 years ago.

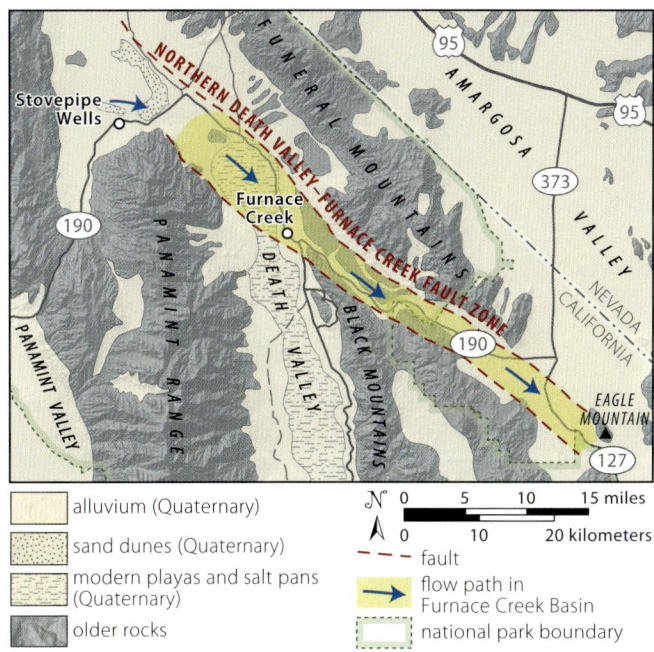

Map of the approximate boundaries of the Furnace Creek Basin showing the direction of sediment transport until the basin became disrupted about 3.5 million years ago. To the northwest, the Furnace Creek fault becomes the Northern Death Valley fault zone. —Arrows from Lutz and others, 2023

In the lower reaches of Furnace Creek Wash, tilted, light-colored sediment of the Furnace Creek Formation is overlain by flat-lying, gray conglomerate of the Funeral Formation, which forms the broad, flat surface in the photograph. Paleozoic rock of the Funeral Mountains forms the skyline.

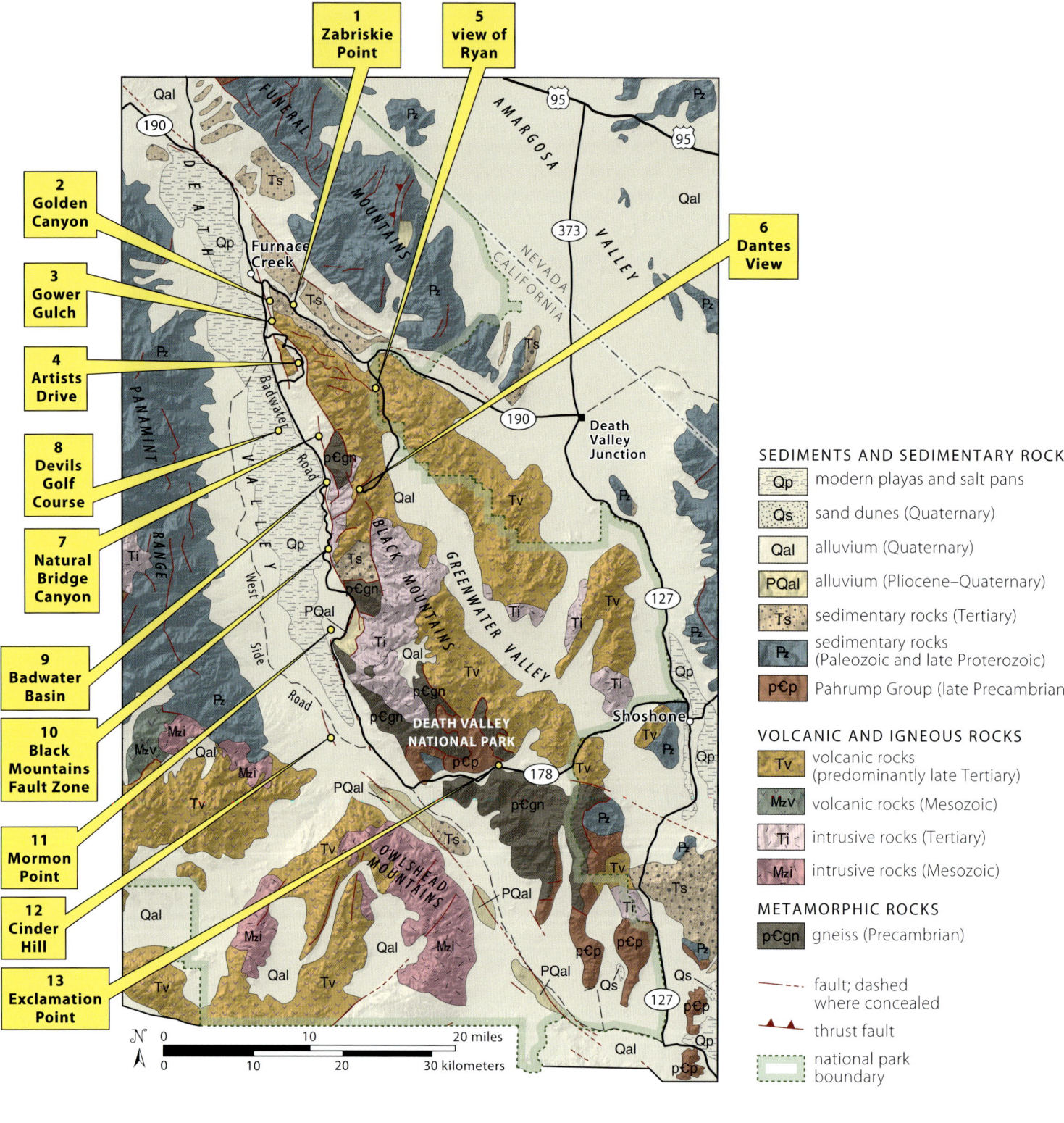

The Black Mountains often look dark because of the dark basement rock that dominates the range south of the Natural Bridge.

BLACK MOUNTAINS AND BADWATER BASIN

The west side of the Black Mountains is one of the most spectacular range fronts of North America. The mountains rise directly from the flat floor of Death Valley, uplifted along the Black Mountains fault zone. Numerous fault scarps and wineglass canyons attest to the fault's recent activity. From the rocks of the Black Mountains, geologists know today's crustal extension began some 16 to 14 million years ago and modern Death Valley's landscape began forming about 3.5 million years ago. Movement on the modern fault will continue in response to the region's ongoing crustal extension.

Unlike the other ranges in Death Valley National Park, which present a striped appearance because of their sedimentary layering, the Black Mountains consist mostly of 11-million-year-old or younger intrusive igneous or volcanic rock and 1.7-billion-year-old metamorphic basement rock. The basement and intrusive igneous rock that dominate the range south of Natural Bridge Canyon tend to present dark somber colors, likely the origin of the range's foreboding name. By contrast, many of the volcanic rocks, as well as the less-common sedimentary rocks, can be startlingly colorful, deriving much of their color through chemical alteration of the rock after it formed.

Surveys of tiny variations in Death Valley's gravitational pull indicate Death Valley is filled with more than 10,000 feet of sediment in places. As the final destination of the Amargosa River and the countless dry stream channels and washes that pour off the mountains, it continually receives silt, sand, gravel, and even boulders. At the same time, the valley drops down along the Black Mountains fault zone, so the sediment gets thicker and thicker through time. Cores of the upper 610 feet of sediment in Badwater Basin indicate that for the last 200,000 years, the sediment thickness has grown by almost 1 millimeter per year. That might seem slow until you account for geologic time: Death Valley began forming about 3.5 million years ago. That's 3,500,000 millimeters—more than 2 miles.

Periodically, Pacific storms drop so much rain that much of the valley floor floods with water, most recently in 2023 and 2005. While the flooded playa renews the cycle of salt dissolution and precipitation, it also reminds us of how responsive Death Valley is to climate. In the past 186,000 years, the valley was filled at least twice by a lake, the largest of which, between 186,000 and 128,000 years ago, reached depths approaching 600 feet. During this time, Death Valley was likely connected to other nearby basins. Evidence of the deepest lake is preserved in the sediment cores and as shoreline features well-above sea level around the margins of the valley.

ZABRISKIE POINT
Stream Capture in the Badlands

One of Death Valley's most popular viewpoints, Zabriskie Point offers a great view of its geology as well as wonderful badlands eroded in rocks of the Furnace Creek Basin. Badlands, like the soft-looking hills below the viewpoint, typically result from a combination of rock type and climate. The rocks are mostly fine-grained sandstones and siltstones deposited in a playa and containing a high percentage of volcanic ash that has been converted to clay. Their fine-grained, clayey nature prevents precipitation from infiltrating the ground surface, which, in combination with the arid climate, results in little vegetation. What rain that does fall typically comes suddenly and runs off quickly, eroding steep gullies. The freshly wetted clays expand, and continual wetting and drying cycles cause the rock at the surface to crack and gradually flow down the slopes, smoothing them.

The tilted layers in the reddish cliffs to the northwest of the parking lot are composed of conglomerate, and the narrow black ridges within the badlands are basaltic lava flows. Together, these rocks are part of the Furnace Creek Formation, deposited between about 6.0 and 2.5 million years ago.

You can inspect an outcrop of fine-grained sandstone and siltstone of the Furnace Creek Formation at the top of the path leading to the viewpoint. You can also walk down the trail into the badlands from the north end of the parking lot to see them interbedded with the overlying conglomerate. If you follow the trail all the way down Golden Canyon (site 2), you will walk into an older conglomerate.

The stream channel in Furnace Creek Wash takes an abrupt westward turn on the southeast side of the parking lot and plunges into a narrow

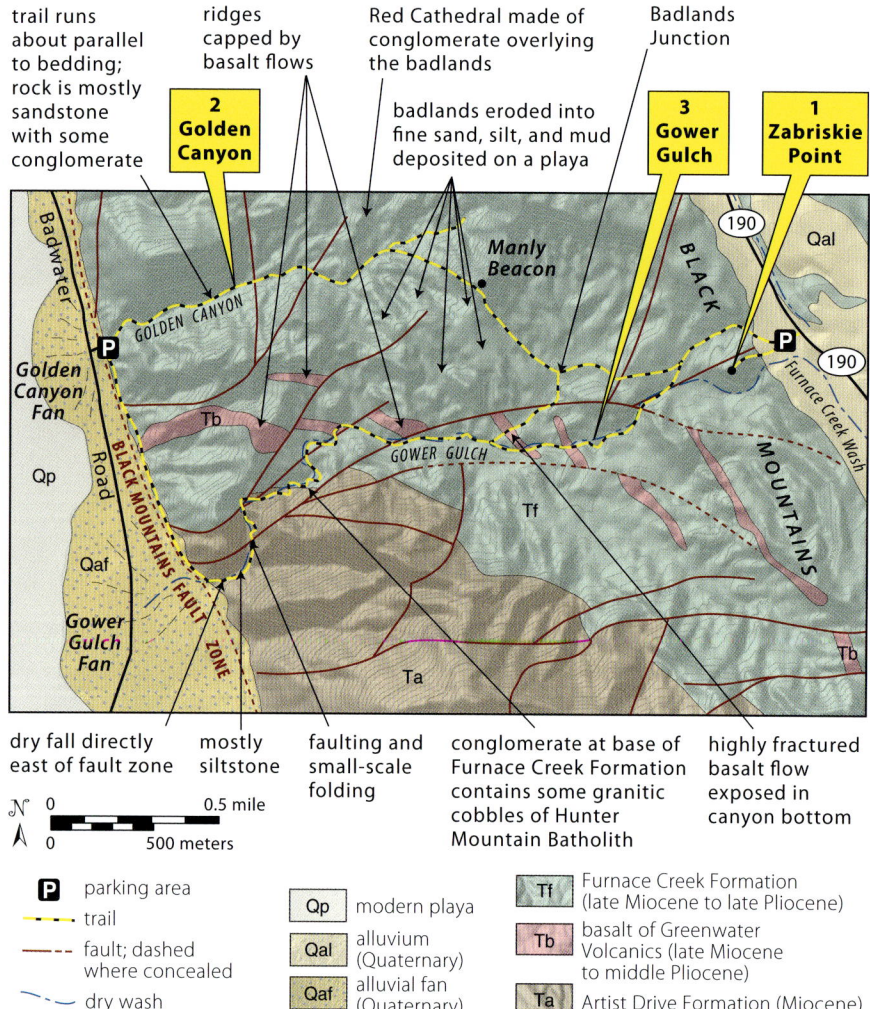

The Zabriskie Point site is 3.5 miles east of the intersection of Badwater Road with CA 190. The parking lot for the Golden Canyon and Gower Gulch Trailheads is directly off Badwater Road, 2.1 miles south of CA 190. —Geology adapted from McCallister, 1970

canyon. In 1941, the wash was diverted into Gower Gulch in an attempt to prevent downstream flooding. This action caused the entire upper part of the watershed—some 165 square miles—to be captured by the tiny Gower Gulch, which previously drained an area of just under 2.25 square miles. The increase in water volume widened the channel in Gower Gulch, deepened the channel near the diversion and at its mouth, and expanded the alluvial fan on the floor of Death Valley. During storms in 1985, 2004, and 2023, floodwaters overwhelmed the diversion and mostly flowed down Furnace Creek Wash anyway.

In the foreground is an outcrop of siltstone at Zabriskie Point overlook. The badlands in the middle ground were deposited in temporary lakes and underlie conglomerate of the red cliffs at right, deposited in streams. The spire in the badlands is Manly Beacon.

The channel in Furnace Creek Wash turns abruptly westward before reaching the Zabriskie Point parking lot and pours down Gower Gulch (to the right of the photo view) because of a human-made diversion, built in 1941. However, the narrowness of the diversion channel allows it to accommodate only a small portion of the water flowing during large floods. View to the southeast.

View to the south with conglomerate of the Furnace Creek Formation at left, the badlands and Zabriskie Point overlook in the middle right, and darker volcanic and sedimentary rocks of the Artist Drive Formation on the crest of the Black Mountains in the distance. Note how the rock layers are all dipping eastward.

GOLDEN CANYON
Walking into the Furnace Creek Basin
See map on page 16.

To hike up Golden Canyon is to experience part of the Furnace Creek Basin, Death Valley's forerunner. The basin extended from the vicinity of Stovepipe Wells southeast to the Amargosa Valley. Rivers carried material into the basin from as far away as the north sides of Saline and Panamint Valleys. Temporary lakes occupied low areas and left behind dry lakebeds, called playas, when they drained.

The canyon mouth at the trailhead is in conglomerates deposited mostly in stream channels. The rocks are difficult to see near the canyon mouth because they're largely coated by dust from windstorms. As you get deeper into the canyon, however, you see gravel of all sizes, ranging from sand to pebbles to boulders, packed together. Most of the gravelly, rounded rocks are volcanic in origin, but some are sedimentary, having come from a variety of localities to the northwest.

The bedding in the rock tilts northward, so where you continue in that direction, you're walking into younger rock. The canyon turns roughly eastward after about 500 feet, after which you walk along much the same group of layers. Some of the sandstone contains ripple marks from running water, and many other beds are green, colored by tiny grains of the green mineral celadonite disbursed through the rock. Celadonite forms by alteration of rocks that are rich in volcanic ash. Look for small faults offsetting the beds. Similar to the main fault along the range front, these normal faults formed from crustal extension.

The canyon turns sharply northeastward after a little more than three-quarters mile, past which are sediments of a playa environment, mostly fine-grained sandstone and thinly layered siltstone. These rocks erode easily and contain clay that expands when wet, which makes them appear unusually soft and rounded. Because no plants are present to slow runoff, the slopes erode quickly into other-worldly badlands. Reddish cliffs of more conglomerate, called the Red Cathedral, loom over the playa beds, signaling a return to gravelly rocks deposited in channels.

After 1 mile, the trail rises into the badlands. It leads to a divide beneath the south face of Manly Beacon and then descends into Gower Gulch, reaching the sign for Badlands Junction at 1.8 miles. From there, continue 1.1 miles up the trail to Zabriskie Point or walk 1.6 miles down Gower Gulch to turn the hike into a loop. If you hike the trail to Zabriskie Point, you'll pass through more badlands until near the top, where you'll encounter increasingly more and more of the gravel. If you hike down Gower Gulch (site 3), you'll see older rocks: the lower part of the Furnace Creek Formation and, below that, rocks of the Artist Drive Formation.

Hiker descending through the badlands toward Gower Gulch.

Ripple marks in sandstone along the south wall of Golden Canyon. Hiking poles for scale.

Hikers walking down Golden Canyon framed by north-tilted beds of mostly sandstone and siltstone of the Furnace Creek Formation.

Mouth of Gower Gulch at the lip of the dry fall that was uplifted by and lies immediately behind the Black Mountains fault zone. The hiker is on the down-dropped side of the fault that lies just off the photo to right.

GOWER GULCH
Taste the Sediment in a Wineglass Canyon
See map on page 16.

Gower Gulch is one of the few wineglass canyons in the Black Mountains you can hike without any technical rock climbing. The trail neatly bypasses the 20-foot dry fall at its base, the steep, narrow part of the canyon that forms the wineglass stem. The trail then enters the main canyon—the wineglass bowl—just past an exposure of the Black Mountains fault zone. Look for a vertical crack in the gully wall about 10 feet in front of (downstream from) the dry fall, which has eroded back from the fault plane. Any future movement on the fault would elevate the range, thus lengthening and steepening the canyon (stem), but the bowl would remain unchanged, although lifted to a higher elevation.

The first quarter mile of trail passes over the upper part of the Artist Drive Formation, which mostly consists of yellowish-orange siltstones and gray limestones near the canyon mouth and conglomerates farther up the canyon. Keep an eye out for small folds in the siltstone and for the many faults of all sizes that cut all the rock types. The canyon turns northward just past a narrow, steep stretch composed of conglomerate and sandstone and after a few hundred yards reaches conglomerate at the base of the Furnace Creek Formation. This conglomerate is markedly thicker and generally contains a greater concentration of volcanic gravels and larger rocks than the Artist Drive conglomerate.

Site continues on page 22 ⟶

The dry fall at the entrance to Gower Gulch. The Black Mountains fault zone, which uplifted the range, is the obvious crack in the gully wall about 20 feet to the right of the dry fall (arrow). In this photo, it looks like it is dipping down to the left, but it is vertical. You can see how it continues through the rock above it.

These layered beds of fine-grained sandstone and siltstone were deposited in a playa of the Furnace Creek Basin. Inset shows crystalline borate deposit found in the canyon walls.

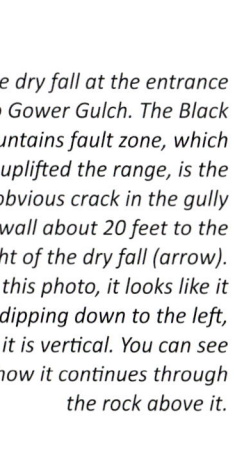

Basal conglomerate of the Furnace Creek Formation (at left) faulted against green and red sandstone of the Artist Drive Formation (at right). Inset photo shows the large cobbles of granite (G) derived from the Hunter Mountain batholith and carried here by rivers. Pocketknife for scale.

Among the many rounded rocks included in this part of the Furnace Creek conglomerate are pieces of granite unambiguously derived from the Hunter Mountain batholith some 35 to 40 miles to the northwest. Similar granite gravels and cobbles exist elsewhere near the base of the Furnace Creek Formation and in river deposits of the 11-million-year-old Eagle Mountain Formation in the Amargosa Valley (site 35). Researchers have used the presence of these granites to reconstruct the river system that fed the Furnace Creek Basin and carried the granite gravels downstream.

Just over a half mile from the canyon mouth, the canyon widens into badlands in fine-grained sandstone and siltstone. Look for light-gray borate deposits in some of the outcrops. The black rocks above the canyon to the north are part of a 6.1-million-year-old basalt flow within the Furnace Creek Formation. As far back as Siegfried Muessig's work in 1954, geologists have considered many of the borate deposits in the Furnace Creek Formation to have resulted from hydrothermal activity that accompanied this volcanism. Some of the basalt flow, highly fragmented from flowing onto a wet playa, shows up at the canyon bottom just shy of Badlands Junction. From there, you can continue through the badlands to Zabriskie Point (site 1) or take the trail down Golden Canyon (site 2).

ARTISTS DRIVE
A Colorful Palette Pleaser

Artists Drive, a 9-mile, one-way road winding through outcrops of the colorful Artist Drive Formation and overlying Funeral Formation, takes you right up to the Black Mountains fault zone. Set your odometer to zero when you begin the drive.

The road climbs an alluvial fan for the first mile before entering a shallow canyon eroded through multicolored lavas and tuffs of the Artist Drive Formation. At 1.4 miles, you can pull off and walk to an overlook that views some 4,000 feet of interbedded volcanic and minor sedimentary rocks of the Artist Drive Formation. Deposited between about 14 and 6 million years ago, they record a major period of volcanic activity during early phases of extension in the Black Mountains.

At about 2.5 miles, a pullout gives a wonderful view of pale-orangish, near-polished, steep fault surfaces along the Black Mountains fault. Striations, or lines, on the surface slant slightly down to the left as you face the cliff. They formed parallel to the fault's slip direction, which was not straight down its plane like a true normal fault, nor parallel to the range front like a strike-slip fault. Instead, it slipped between the two, in an oblique motion approximately parallel to the regional direction of extension, which is northwestward.

At 2.9 and 3.4 miles, the road dips into wineglass canyons that formed from recent movement along the Black Mountains fault zone. Their steep narrow parts, which you can see if you look

The Black Mountains fault zone (yellow arrow) traces a straight line just to the right of Artists Drive at lower right. At mile 3.4, the road crosses a well-defined wineglass canyon (WC). Artists Palette (AP) lies just below and left of the photograph's center. Furnace Creek Wash is the broad area behind the crest of the Black Mountains. Behind them are the Funeral Mountains.

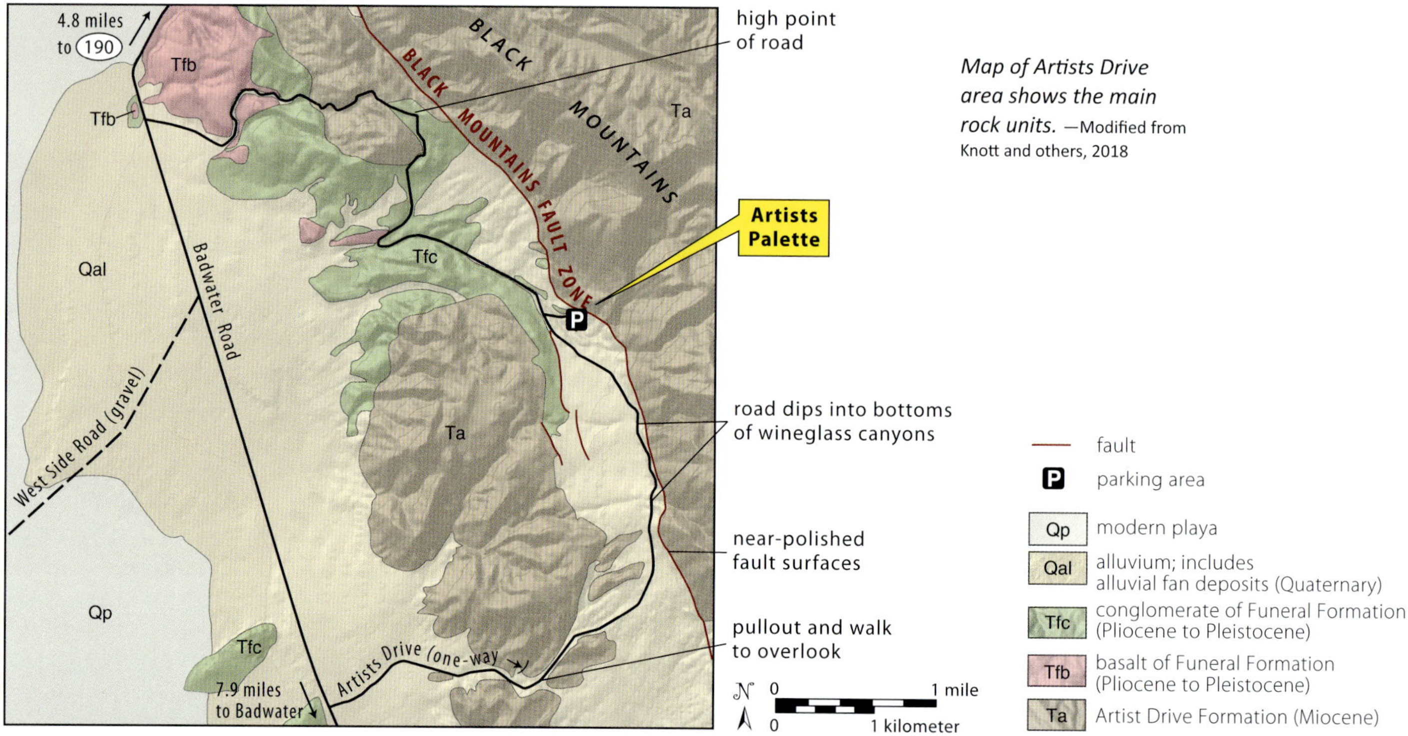

Map of Artists Drive area shows the main rock units. —Modified from Knott and others, 2018

—	fault
🅿	parking area
Qp	modern playa
Qal	alluvium; includes alluvial fan deposits (Quaternary)
Tfc	conglomerate of Funeral Formation (Pliocene to Pleistocene)
Tfb	basalt of Funeral Formation (Pliocene to Pleistocene)
Ta	Artist Drive Formation (Miocene)

Artist Drive Formation forms the low hills in the foreground and high cliffs in the background as seen from the high point along the road at mile 6.9.

Perhaps the most-striking color at Artists Palette is seafoam green, caused by the mineral celadonite.

upstream from the road crossings, lie just east (uphill) of the fault. Recent uplift on the fault steepens the mountain front, so the narrow canyon has had little time to erode. The bowl area, some distance behind the fault, continues to widen and become less steep through erosion.

The turn-off for Artists Palette is at 4.3 miles. The colorful rocks consist of basalt and rhyolite lava flows and breccias with ash-flow tuffs of the Artist Drive Formation. The rock compositions, as well as their later hydrothermal alteration, produced the variety of colors. Blacks, browns, reds, and pinks typically come from oxidized iron, whereas the greens likely come from the mineral celadonite, which contains both oxidized and reduced iron and forms by alteration of ash-rich tuffs.

The road descends to a hairpin turn at 5.9 miles at an outcrop of steeply dipping conglomerate of the Funeral Formation and then heads back uphill toward the Black Mountains fault zone, passing more alluvial fan deposits and fractured basalts of the Funeral Formation. Upon reaching the highpoint near mile 6.9, you can see green and white deposits of the Artist Drive Formation a short way off the right side of the road. Beyond the high point, the road descends through Artist Drive Formation from about 7.2 to 7.8 miles and then through alluvial fan deposits of the Funeral Formation. At 8 miles, it enters basalt, which it follows until the canyon ends on the modern alluvial fan.

VIEW OF RYAN
Mining Camp That Extracted Borate
See map on page 27.

Once home to miners who worked a world-class borate deposit, the mining camp of Ryan sits perched atop cone-shaped talus deposits of black basaltic boulders derived from the lava flows above. You can get a good view of Ryan about 2.5 miles south of CA 190 along Furnace Creek Wash Road, the route to Dantes View. From the highway, Ryan looks tiny, but it was a company town in the 1920s operated by the Pacific Coast Borax Company. In addition to living quarters, the town hosted a post office, hospital, and school.

Named Lila C when it was built in 1907 to support the nearby Lila C Mine, the town was renamed Ryan to honor the company's general manager who died in 1918. As more borate mines in the area opened, Ryan became the western terminus of the Death Valley Railroad and the hub for a network of narrow-gauge tracks that linked the mines. As the region's greatest producer of borate minerals, it shipped out more than $30 million worth of ore before outside competition forced it to close in 1928. It lies just outside the national park and is presently owned by the Death Valley Conservancy, which leads occasional tours of the site.

Borate minerals are widely used in fertilizers, insecticides, detergents, and even as fluxes or hardeners in metal manufacturing. The numerous underground mines at Ryan extracted mostly the borate minerals colemanite and ulexite from local ore concentrations in the Furnace Creek Formation.

About a half mile to the northeast, you can also see the headframe of the Billie Mine, a large underground mine that extracted the same suite of minerals—also from the Furnace Creek Formation. Although its surface workings lie just outside the park boundary, the mine produced ore from underground that lay inside the park boundary, until it closed for good in 2005.

If you look directly above the middle talus cone behind Ryan, you can see where a near-vertical fault separates tilted, greenish beds of the Furnace Creek Formation on the right from reddish, greenish, and orangish beds of the Artist Drive Formation on the left. Unfaulted, 4-million-year-old basalt of the Greenwater Range overlies these rocks. This important relationship indicates major faulting in this part of the Furnace Creek Basin ended by the time the basalt erupted at 4 million years ago.

View of Ryan from the road to Dantes View. The middle cone of black talus covers a fault that separates the Artist Drive Formation on the left from greenish Furnace Creek Formation on the right. The overlying basalt flows, which have been dated at 4 million years old, are not faulted.

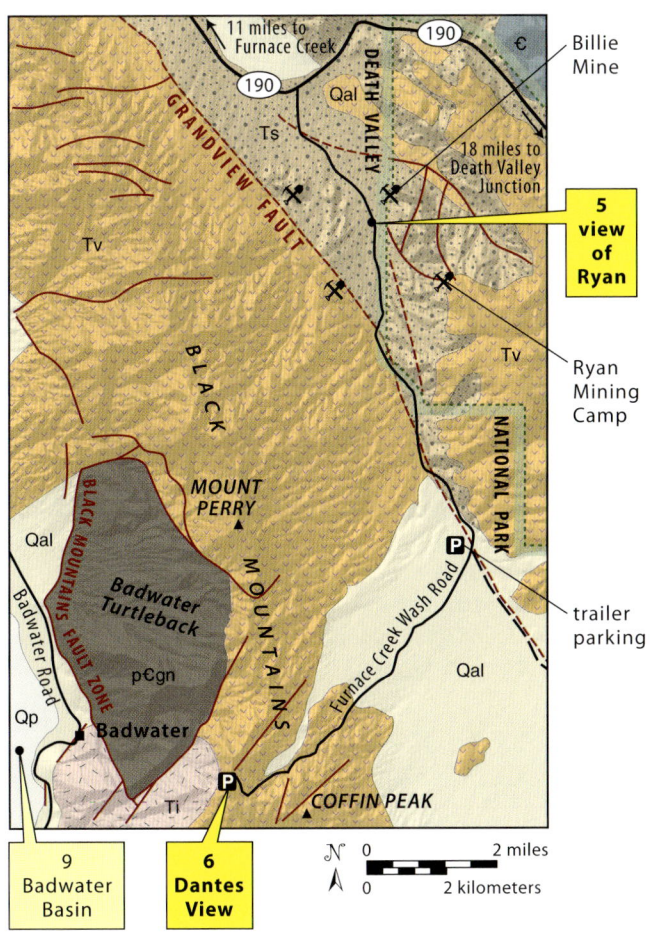

DANTES VIEW
One Mile Straight Down

Nothing on the drive to Dantes View prepares you for the experience of pulling into the parking lot and looking from the top of the Black Mountains down into Death Valley. In contrast to the relatively gentle grade the road climbs on the east side, everything drops off suddenly to the west. Badwater, with its wide trail out to the salt pan, is visible more than 1 mile below. This asymmetry exists because the range is a tilted fault block uplifted along the Black Mountains fault zone. The mountains rise steeply behind the fault and tilt gently eastward. Looking both east and west, you can see more ranges, also tilted fault blocks, nearly parallel to each other and uplifted by normal faults responding to crustal extension.

Directly across Death Valley, the Panamint Range forms another tilted fault block. Like the Black Mountains, they rise along a normal fault along their west side and tilt eastward, reaching an elevation of 11,049 feet at Telescope Peak (site 24). Because there is little recent faulting along the east side of the Panamints, erosion has had plenty of time to eat away at its shape, resulting in a highly irregular range front.

The gentle eastward tilting also affects the alluvial fans and salt pan on the floor of Death Valley. Those fans on the west (Panamint) side are able to spread out and coalesce into giant bajadas because the gravel moves down a long, steady grade, whereas those on the east (Black Mountains) side can't spread very far because the valley floor keeps dropping along the fault and tilting eastward. The Panamint fans are also bigger because the higher, wetter Panamint Range is home to bigger drainage basins and faster rates of erosion.

The eastward tilting also affects the valley floor. It looks perfectly flat, but notice how the salt lies a lot closer to the east side of the valley than the west. Imagine a shallow bowl half full of salty water. If you were to set it on a table and let the water evaporate, the salt would precipitate as concentric rings, with the

Ryan and the Billie Mine headframe are clearly visible from the road leading to Dantes View, about 2 to 3 miles up the road. Dantes View lies at the very end of the road, 13 miles from CA 190.

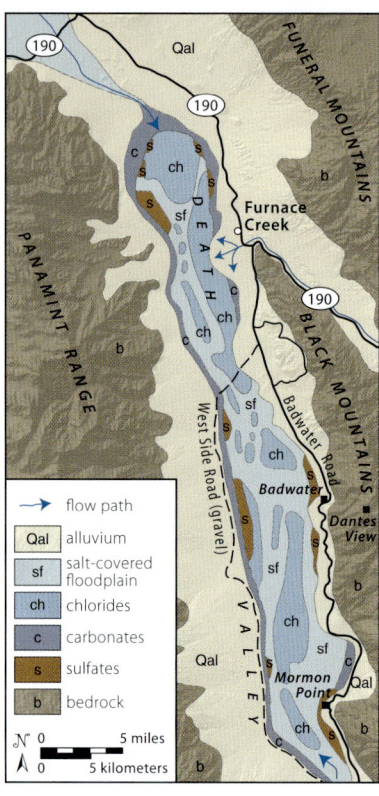

Map showing the asymmetric zonation of the salt pan, with the least-soluble carbonate zone along the edges and better developed on the west side, with more-soluble sulfate zone interior to that, followed by the most-soluble chloride zone. Each zone contains numerous minerals of that type, but the best known of each include travertine (carbonates), gypsum (sulfates), and halite (chlorides).
—Modified from Hunt, 1975

less-soluble salts precipitating in the outer rings and the more-soluble ones, like halite (table salt), in the center. If you were to tilt the bowl and let it evaporate, the rings would be skewed, with wider zones of precipitation on the upslope side and a concentration of the more-soluble salts on the downslope side. Such asymmetric zoning is exactly what Charles Hunt of the US Geological Survey discovered in the 1960s and 1970s, indicating the valley floor continually tilts down to the east as crustal extension proceeds.

The rocks at Dantes View are all volcanic, with ages of about 6 million years. You can get a good look at outcrops of ash-flow tuff if you walk up the trail to the north. These tan-colored rocks formed during explosive eruptions, as indicated by numerous pieces of pumice as well as rock fragments included in the matrix of hard, welded ash. These volcanic rocks rest on older metamorphic and intrusive igneous rocks of the crystalline basement that were metamorphosed about 1.7 billion years ago. To the north, you can see greenish gneiss of the Badwater Turtleback, a mass of Precambrian rock with a surface that resembles a turtle shell. Like the Copper Canyon and Mormon Point Turtlebacks to the south, it was buried to depths greater than 10 miles and then uplifted along a detachment fault before the most-recent uplift along the Black Mountains fault zone (see sites 7 and 10).

View northward to the Artists Drive area and Furnace Creek alluvial fan. Note the abrupt, fault-controlled western edge of the Black Mountains.

Ash-flow tuff along the trail just north of Dantes View. It consists mostly of volcanic ash with rock fragments picked up by the ash flow.

Partially flooded Badwater Basin as seen from Dantes View in November 2023. The flooding occurred after heavy rains from Hurricane Hilary in August 2023. Note the big alluvial fans extending outward from the Panamint Range. The greenish, sloping rocks in the foreground are gneisses of the Badwater Turtleback.

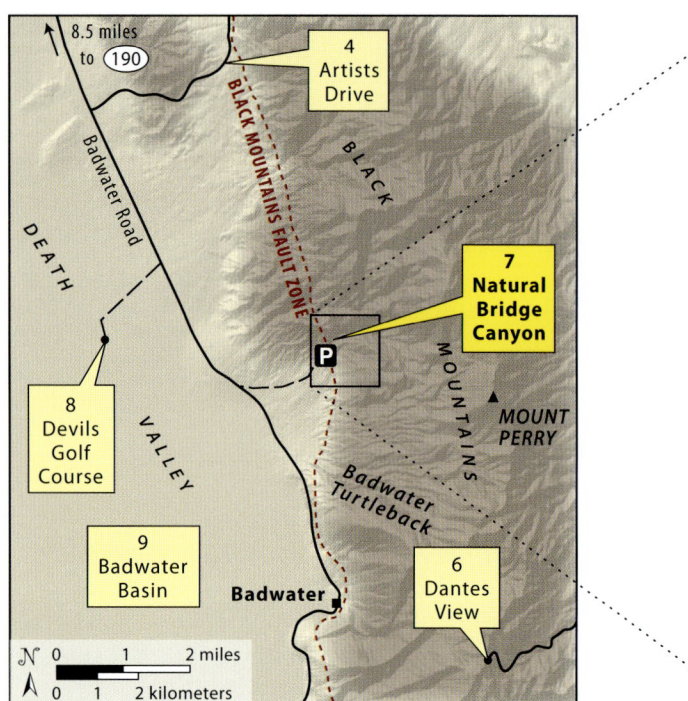
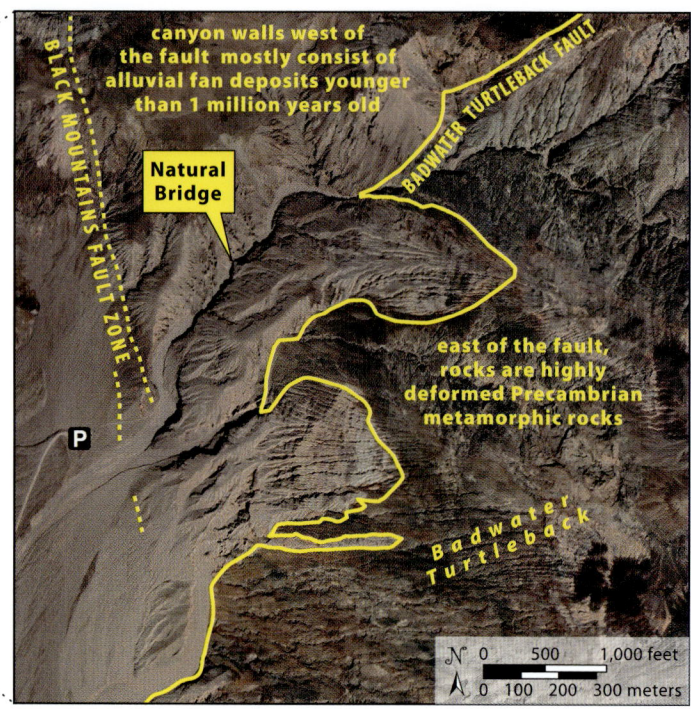

The trailhead for Natural Bridge Canyon starts at the end of a fairly rough 1.5-mile-long gravel road leading from Badwater Road, 13 miles south of its intersection with CA 190.

Satellite view of Natural Bridge Canyon, showing location of the natural bridge, the turtleback fault zone, and the trace of the Black Mountains fault zone. —Modified from a Google Earth satellite image, 2024

NATURAL BRIDGE CANYON
Fantastic Fanglomerate Faulted over Gnarly Gneiss

When you hike up Natural Bridge Canyon, you get an inside view of an alluvial fan that formed a bridge and is faulted against basement gneiss. The walk to the natural bridge takes only about 15 minutes, and you'll cross a fault scarp in the loose alluvium just before the canyon mouth. From there, you walk up a steady incline of about 6 degrees on the gravel-covered canyon bottom. The loose gravel, eroded from the canyon walls and upper reaches of Natural Bridge Canyon, now gradually makes its way down to the modern fan.

Alluvial fan deposits, known as fanglomerate, form the canyon walls. They contain the 772,000-year-old Bishop ash bed, which provides their age and tells us they are part of the Mormon Point Formation that you can also see at site 11. The deposits seem unlikely candidates to hold up a natural bridge, but clues next to the bridge speak to its origin. At the same level as the bottom of the bridge's arch, dark stripes on both canyon walls mark positions of water seepage and likely coincide with the canyon bottom at an earlier time. An arcuate-shaped cavern, a former meander bend, at this same level wraps around the north side of the bridge. Before the canyon eroded to its present level, water flowed around the meander bend, bypassing the wall that became today's bridge. In the process, the water eroded its way through the bottom of the

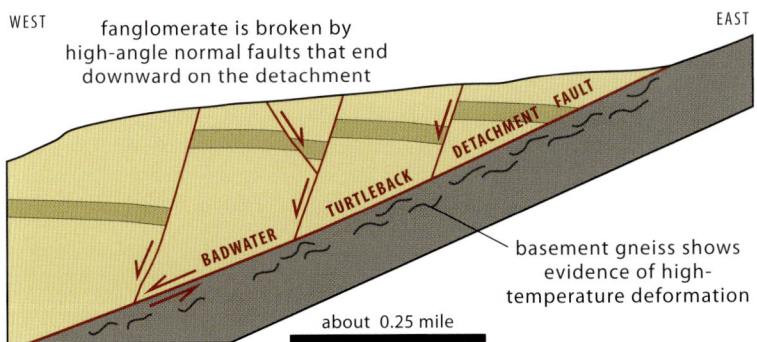

Schematic cross section showing the relationship of high-angle faults in the fanglomerate of Natural Bridge Canyon to the Badwater Turtleback detachment fault.

View southward along the front of the Badwater Turtleback. The turtleback fault lies between the Precambrian metamorphic rock and much younger Cenozoic volcanic rock and fanglomerate along its base. The fault runs along the entire front, but it is labeled in only one place. The Black Mountains fault zone runs along the base of the hills, uplifting the fanglomerate and volcanic rock above the modern fan.

wall, creating a small window to allow it to flow down the canyon in a straighter line. As the canyon eroded more deeply, the new channel passed beneath the wall, creating the bridge.

The canyon walls are cut by myriad normal faults, attesting to the modern crustal extension. In some places, vertical chimneys formed where water poured over the canyon edge. Just past the bridge, a giant alcove formed where water flowing down the canyon channel eroded into the canyon's side, much like in the meander bend by the bridge.

A quarter mile beyond Natural Bridge, the canyon bottom abruptly changes from gravel to bedrock at a small dry fall. Here, the fanglomerate is juxtaposed with highly deformed metamorphic rock along the Badwater Turtleback fault, one of Death Valley's detachment faults. A reddish zone of fine-grained material, called fault gouge, marks the fault and formed through the grinding and breaking of the rock during fault movement. The fault continues westward underground, likely serving as the bottom of many of the higher-angle faults in the canyon. The same fault zone is also visible looking southward from the trailhead parking lot, separating the low reddish hills of fanglomerate and volcanic rock from the greenish to brown metamorphic rock.

Upstream from the dry fall, you can get a close look at the metamorphic basement rock. The canyon walls become narrower as you encounter the brown and gray marble and green gneiss, both

Badwater Turtleback fault, the reddish zone (behind the author) that diagonals across the photo, separates the fanglomerate (above) from highly deformed Precambrian metamorphic rock (below). Because of its low angle, the fault is exposed on both sides of the canyon. The yello arrow on the right side of the photo also points to the fault. The gravel below the author is not part of the fanglomerate but a more-recent canyon deposit.

Natural Bridge spanning the canyon walls with the stranded meander bend showing in the sunlight behind and to the left of the bridge. The alluvial fan deposits in the lower canyon walls consist exclusively of volcanic rock.

of which contain abundant white bodies of pegmatite, igneous intrusions with large crystals. The linear pattern on the gneiss surfaces formed parallel to northwest-directed shearing when the rock was deformed at high temperatures. This high-temperature deformation characterizes the Badwater Turtleback, the dome-like mass of metamorphic rock that forms the mountain front from here to Badwater.

Unlike the modern alluvium, which consists of both volcanic and metamorphic gravels, the fanglomerate you can see along the lower canyon walls contains only volcanic pieces. The absence of metamorphic rocks indicates that unlike today, no metamorphic rocks were exposed when these deposits accumulated. Near the top of the canyon walls, however, in an undated younger part of the fanglomerate, you can find metamorphic cobbles, marking the time when those rocks were first exposed and eroded.

View looking east to the Badwater Turtleback from Devils Golf Course (site 8), showing its convex-upward shape and large wineglass canyon. Natural Bridge Canyon marks the northern edge of the turtleback.

Near the entrance to Natural Bridge Canyon is a small fault scarp of the Black Mountains fault zone. Arrow points to its bottom.

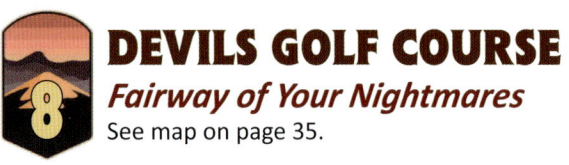

DEVILS GOLF COURSE
Fairway of Your Nightmares
See map on page 35.

When the salt pan floods, parts of it manage to remain dry, simply because they're a little bit higher. These areas coincide with the crests of gentle folds in the salt pan and likely reflect modern-day tectonic activity. Devils Golf Course, a slightly higher part of the salt pan that seldom floods, appears to lie on the crest of one of these folds. It did not flood during the extreme events of 2005 and 2023. The effects of wind and dissolution from rainwater tend to predominate over those of new salt precipitation. They fashion the uplifted polygons into a landscape of pedestals rimmed by sharp, tooth-like salt decorations. Notice these decorations are mostly oriented north-south, parallel to the prevailing winds. Some of the pedestals reach heights of 3 feet above their bases.

From Devils Golf Course, you gain an outstanding view of the Badwater Turtleback in its entirety on the face of the Black Mountains. Its convex-upward shape, which to some people resembles the shape of a turtle shell, resulted from folding of the metamorphic rock. The turtleback, which forms its northern edge as the deep canyon curves gently to the southeast around the fold, is described in Natural Bridge Canyon (site 7). North of the canyon are the colorful volcanic rocks of the Artist Drive Formation that are faulted against the metamorphic rock.

Close-up view of needle-like decorations formed in the salt at Devils Golf Course. Photo is 5 inches across at the bottom.

View to the south across the wind- and rain-eroded salt pinnacles at Devils Golf Course. Note how the pinnacles are oriented north-south, parallel to the prevailing winds.

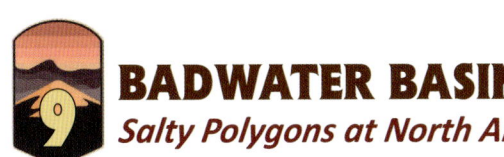

BADWATER BASIN
Salty Polygons at North America's Low Spot

Badwater, at an elevation of 282 feet below sea level, is the lowest spot in Death Valley and in North America. Like many other valleys in the Basin and Range, Death Valley's vast floor is a sink. It receives water, salts, and sediment from nearby sources—the Amargosa River to the south, Salt Creek to the north, and myriad small springs and seeps along its edges—but without an outlet, it releases the water only through evaporation. Much of the valley floor is covered in salt and other alkali minerals left behind by evaporation, along with the ever-present mud, silt, and sand.

A spring at Badwater produces enough water to form a small pond that exists throughout the year. Freshwater issues from the fault zone at the base of the Black Mountains and quickly becomes salty as it passes through the salt pan at the surface. You can see the fault if you look southward a quarter mile from the pond to the alluvial fan. The fault's most-recent movement broke the fan's surface to form a series of fault scarps, small steps in the otherwise smoothly sloping surface.

Perhaps the most-striking features of the salt pan are its expansive stretches of polygons, confined by narrow ridges of salt that intersect at angles of about 120 degrees. You can see these polygons by walking a half mile into the salt pan along the wide path worn by countless visitors. Similar to damp mud, the surface of the salt pan shrinks and cracks as it dries to form cracks that outline a variety of polygonal shapes. As evaporation proceeds, water moves from below the surface upward through the cracks by capillary action and precipitates salt. As the process continues, more and more salt precipitates at the site of the crack, accumulating into a ridge of salt, which continually grows upward. Even a casual

The parking lot for Badwater is along Badwater Road, 16.5 miles south of its intersection with CA 190. Devils Golf Course is 1.3 miles down a dirt road that meets Badwater Road about 11 miles south of CA 190.

Crystallizing salt in shallow water. Note the cubic shapes of the crystals.

Fault scarps, the ridges of gravel, cut the alluvial fan directly to the south of Badwater.

Raised zones of salt that collect dark-colored sediment mark the polygonal fractures on the salt pan. View to the south.

Shoreline-deposited gravel (g) from Lake Manly adorn the side of the Black Mountains behind Badwater and to the north. The bedrock of the mountains visible in this photo consists largely of Precambrian crystalline basement rock (pЄb) and marble (pЄm). Note the dike that cuts across the marble.

inspection of the polygons reveals freshly precipitated salt along these ridges.

On the walk onto the salt pan, you can see how varying amounts of mud, clay, sand, and even gravel, close to the mountain front, also cover the valley floor. These same materials compose the valley fill, as demonstrated by cores drilled to depths of 1,000 feet in the early 1900s by the Pacific Coast Borax Company. Some of the sediment was deposited on the bottom of lakes. You can see shoreline deposits from Lake Manly, one of the most-recent lakes, clinging to the mountain front at an elevation of about 300 feet northeast (left) of the parking area. These deposits consist of gravel cemented with tufa, calcium carbonate precipitated from the lake water. From just a short distance out on the salt pan, you look directly upward toward Dantes View (site 6) and the steep southern end of the Badwater Turtleback (site 7) rising behind the Black Mountains fault zone.

10 BLACK MOUNTAINS FAULT ZONE
Broken and Rebroken Breccia

South of Badwater Spring, Badwater Road circles around the edges of four alluvial fans before coming right up against the mountain face. Small triangular facets of fanglomerate lie faulted against older rock along the Black Mountains fault zone. Look for a red-and-white-striped outcrop as you approach the mountain from the north. The Black Mountains fault zone, where you can see broken and rebroken rock with amazing color changes and patterns, lies behind this outcrop. Be careful if you climb around on it because it can be unexpectedly treacherous.

The rock at this site, which was originally volcanic, is so thoroughly shattered that it is now breccia, a different rock type altogether. Breccias can form in a variety of ways, but they all consist of angular particles. The crude layering in the deposit suggests it might have originated as a sedimentary deposit, but its location along the Black Mountains fault zone suggests it formed by faulting and crushing of the original volcanic rock. The breccia then became altered by circulating groundwater, forming color banding.

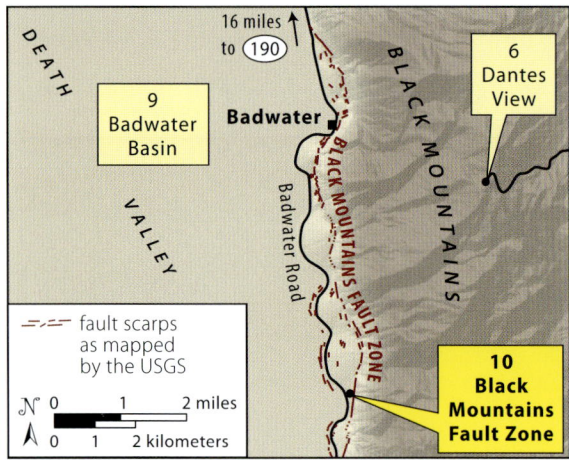

The Black Mountains fault zone extends along the entire front of the Black Mountains. The red lines show places where faults break through the alluvium, forming fault scarps. At this site, 4.6 miles south of Badwater (36°10'38.55" N, 116°45'47.82"W), you can see the effects of the fault exposed in the bedrock.

Fanglomerate faulted against bedrock that is pervasively brecciated and shows color banding from later chemical alteration. Movement along the fault created shiny, smooth, and striated surfaces (close-up).

Faulting of the breccia and color banding attests to repeated movement events along the fault zone. The main faults in the photo slant down to the right in the colorful zone below and including the pocketknife.

The color banding in the rock is disrupted by narrow fault zones nearly parallel to the banding. Look closely to see small faults cutting directly across the color banding.

Thousands of discrete faulting events uplifted the Black Mountains to their present elevations. Remember the Black Mountains rise to elevations well-over 5,000 feet, and the bottom of the sediment that fills Death Valley is more than 10,000 feet deep.

MORMON POINT
Turtles All the Way Across

Similar to Badwater (site 9), spring water issues from the fault zone that separates the mountains at Mormon Point from the valley floor. But Mormon Point juts northward into the salt pan, a product of complicated faulting that created an equally complicated mountain front. South of Mormon Point, the range is long and straight, whereas immediately north of the point, you can see how the range changes its trend abruptly several times, creating a large embayment. As described by Terry Pavlis and his colleagues in the 1990s, several differently oriented faults along the edge of the mountains create this unusual geometry. Moreover, they likely played—and may still play—a role in controlling the earthquake cycle on the Black Mountains fault zone. You can also see a large fault scarp breaking through the gravels a short distance to the northeast of Mormon Point, indicating how recent the fault movements have been.

You get a striking view of the Copper Canyon Turtleback by looking northeast across the embayment. The

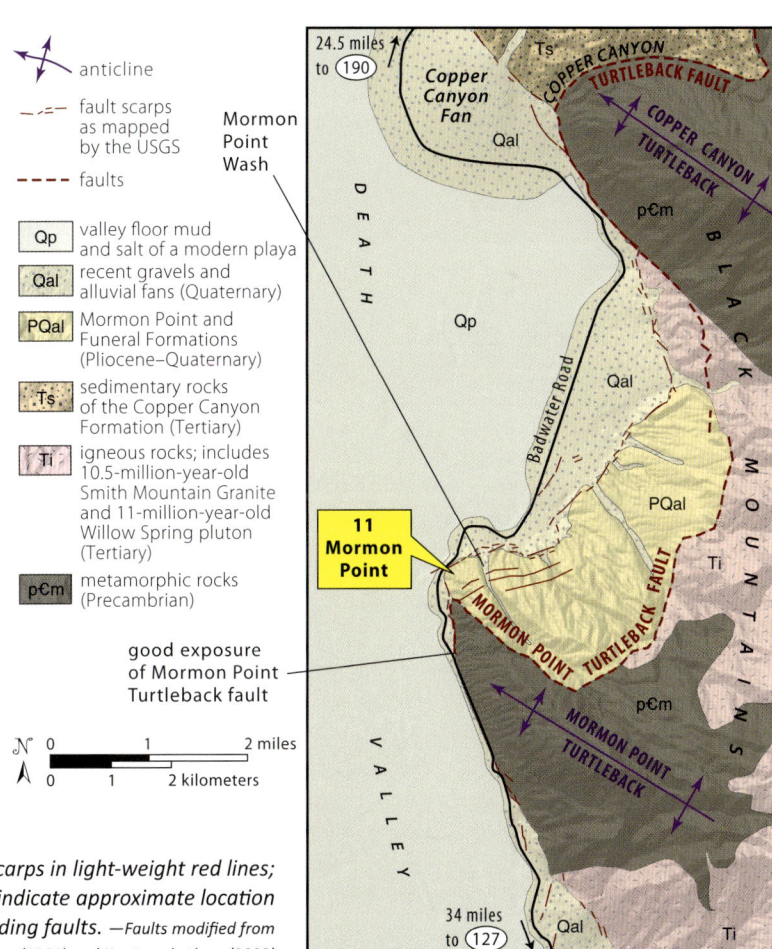

Map shows recent fault scarps in light-weight red lines; heavier dashed lines indicate approximate location of older bedrock-bounding faults. —Faults modified from Brogan and others (1991) and Knott and others (2002)

View of Copper Canyon Turtleback across ponded spring water and salt. The arrow marks the turtleback detachment fault, which angles up to the right, separating the reddish-brown sedimentary rocks of the Copper Canyon Formation from the greenish metamorphic rock of the Copper Canyon Turtleback. High-angle faults in the Copper Canyon Formation end downward at the detachment in a way similar to those in Natural Bridge Canyon (site 7). The green color in the rocks comes from the mineral chlorite, which formed during later, shallower stages of metamorphism as the rocks were being uplifted.

Aerial view of Mormon Point showing Mormon Point Wash, its alluvial fan, and its fault scarps. The star marks a former shoreline at an elevation of 295 feet above sea level. The benches below the star are the same shoreline but faulted downward as determined by Jeff Knott and colleagues in 2002.

Exposure of the Mormon Point Turtleback detachment fault in the small canyon 0.8 mile south of the sign for Mormon Point. The top of the dry fall lies just below the fault, the obvious line slanting diagonally from the upper right to the middle left.

turtleback consists mostly of greenish Precambrian basement gneiss and marble that was strongly deformed at high temperatures along an extensional fault zone that reached deep into the crust. With a little imagination, you can see how the curved shape of the metamorphic rock resembles the back of a turtle, which prompted Donald Curry to start calling them turtlebacks in 1938. The turtleback rocks are faulted against overlying reddish and tan sedimentary rocks that belong to the 5- to 3-million-year-old Copper Canyon Formation. They were deposited in a small basin at about the same time that much of the Furnace Creek Formation was being deposited in the Furnace Creek Basin.

At Mormon Point, you stand on the crest of the Mormon Point Turtleback, but it's difficult to see any of its features from the road. Both it and the Badwater Turtleback, which lies some 10 miles to the north, resemble the Copper Canyon Turtleback in that they consist of deformed metamorphic rock faulted against sedimentary or volcanic rock and they have a dome-like, turtle shape. You can access the Mormon Point Turtleback detachment fault and the metamorphic rocks beneath it by hiking 1.5 miles up Mormon Point Wash. Like the hike up Natural Bridge Canyon to the Badwater Turtleback (site 7), you get a good look at the gravel walls, called the Mormon Point Formation. The formation contains ash beds that range in age from 500,000 years to perhaps as old as 1.2 million years. The gravels, cut by numerous faults that likely end where they encounter the detachment fault, display interesting erosional features. You can also access the fault by driving 0.8 mile south of the sign for Mormon Point and walking a hundred yards or so up the canyon.

The Mormon Point Turtleback plunges northwestward below the valley floor, separating two parts of the valley that are filled to great depths by gravel, sand, mud, and salt. Using measurements of tiny variations in the gravitational pull from place to place, researchers have determined Badwater Basin to the north is some 11,500 feet deep while the Mormon Point Basin to the south is probably 10,000 feet deep. Between them, a much shallower zone lies on trend with the turtleback.

12 CINDER HILL
Implications of an Offsetting Volcanic Cone

Two low red hills on the valley floor about 8 miles south of Mormon Point don't look like much, until you realize they formed from erupting magma sometime after Lake Manly occupied Death Valley during the ice age. You can visit the hills by driving 1.6 miles north on West Side Road, which intersects Badwater Road 1.75 miles northwest of Ashford Mill. The red hills are composed almost entirely of loose basaltic rock fragments and red cinders, the color of which is derived from oxidized iron. Cinders are small pieces of basaltic lava that cooled and became full of air pockets from being exploded out of a volcanic vent. They tend to accumulate around the vent to build a small volcano called a cinder cone. These two hills once formed a single cinder cone that was split in two by the Southern Death Valley fault zone. Movement along the fault offset either side of the cone by 699 feet in a right-lateral sense.

It's difficult to pinpoint just when the cinder cone erupted. The cinders overlie 180,000- to 126,000-year-old deposits from Lake Manly and underlie 26,000- to 6,000-year-old alluvial fan deposits about 1 mile to the northeast. An age somewhere in between makes it the youngest volcanic activity in Death Valley aside from the Ubehebe Crater volcanic field (site 26).

Using the oldest age estimate, the rate of slip on the fault zone is just over 1 millimeter per year; using the younger age estimate for the cinder cone, the rate is thirty times faster. A variety of estimates exist on the fault, and the preferred rate lies somewhere between 1 and 8 millimeters per year. For comparison, the US Geological Survey estimates the slip rate of the San Andreas fault in northern California to be about 20 millimeters per year.

In the mid-1980s, Beatrice de Voogd and her colleagues used data from seismic reflection experiments to image a zone some 10 miles beneath the cone that was likely a magma body—the possible source of the volcano. They also found evidence of a normal fault that connected the zone to the cinder cone and might have acted as a conduit for the magma to travel to the surface. Scattered about on the cone surface are pieces of light-gray sandstone welded to the basalt, another indication of what lies underneath.

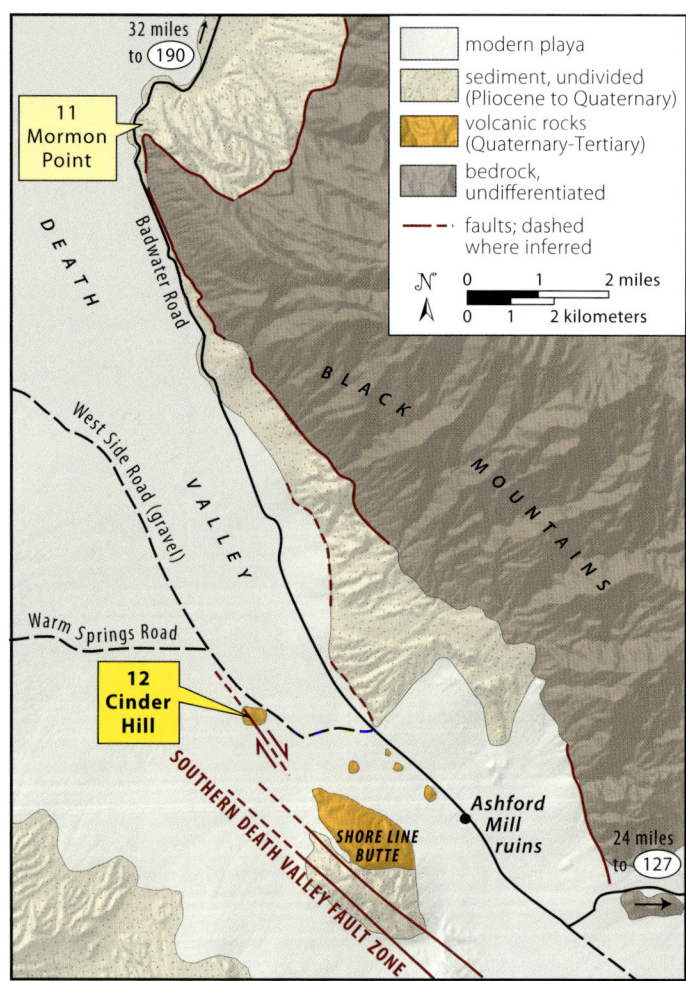

Cinder Hill, offset by the Southern Death Valley fault zone, sits just south of West Side Road, 1.6 miles from its intersection with Badwater Road, 27 miles south of Badwater.

View looking north over Cinder Hill. The Southern Death Valley fault zone, which runs down the valley between the two hills, offset the once-single cinder cone. In the background, the relatively gentle Panamint Range on the left and steep Black Mountains on the right display Death Valley's asymmetry. Inset shows a single cinder with holes from gas bubbles.

This 5-inch-long sandstone fragment, rimmed by basaltic lava, was brought to the surface during the eruption of Cinder Hill.

Cinder Hill, as seen from Badwater Road, was offset to the right by the Southern Death Valley fault zone.

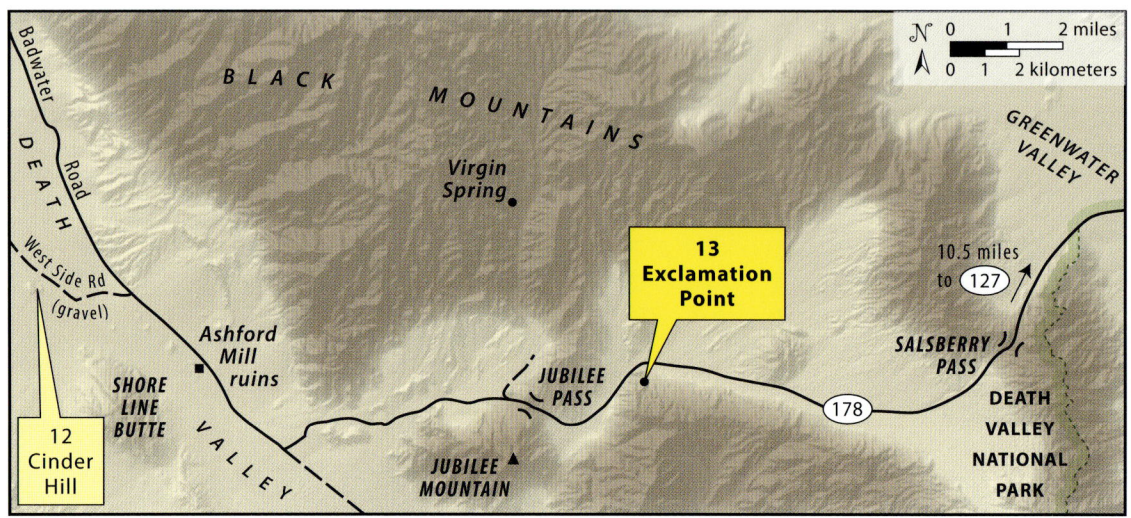

Exclamation Point lies just 2.5 miles east of Jubilee Pass.

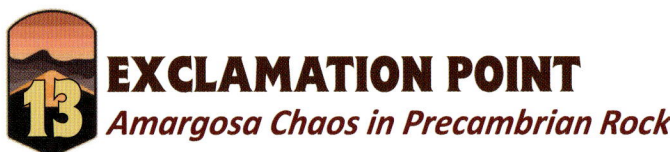

EXCLAMATION POINT
Amargosa Chaos in Precambrian Rock

On the south side of CA 178 and just 2.5 miles east of Jubilee Pass, Exclamation Point provides one of the best exposures of the Amargosa Chaos, an aptly named area defined by intensive faulting. Lauren Wright and Bennie Troxel, who mapped much of the Death Valley area over a period spanning six decades, called this outcrop Exclamation Point because of its unusual ability to elicit happy cries of epiphany from other geologists.

In the 1930s, Wright and Troxel's mentor, Levi Noble, was the first person to study the Chaos in detail. Wright and Troxel published a detailed map of the area in 1984, but we are still trying to understand many of the area's nuances. Wright and Troxel found that most of the faults are normal or strike-slip and are related to crustal extension, most of which is older than 10 million years. They also noted the contact between the sedimentary rocks and the crystalline basement was typically—but not everywhere—faulted. In some places, the lowest sedimentary rock was in its original depositional position on the basement. Later researchers have found that, where the contact is faulted, its direction of movement varies.

At this exposure, which is entirely Precambrian rock, you can see the crystalline basement at the bottom of the outcrop and successively younger rock exposed upward, each appearing as an individual fault sliver. A dark-reddish ledge of dolomite of the Crystal Spring Formation, for example, is faulted against the basement, and above that are multiple slivers of diabase and overlying Noonday Dolomite. At the very top of the outcrop are slivers of the upper Johnnie Formation. Considering the Crystal Spring Formation and Noonday Dolomite have an original thickness of about 1,200 feet, the faults in this outcrop cut out a significant amount of rock. A small avalanche deposit covers many of the features near the center of the exposure. If you walk around the back of the exposure, you find more crystalline basement.

Close-up view of the whitish fault zone between the crystalline basement rock (bottom) and the Crystal Spring Formation (top). The pencil rests on top of the basement. Note the high-angle crack—another fault—in the Crystal Spring Formation that ends downward into the fault zone.

A close-up of the eastern end of the outcrop showing faults (dashed yellow lines).

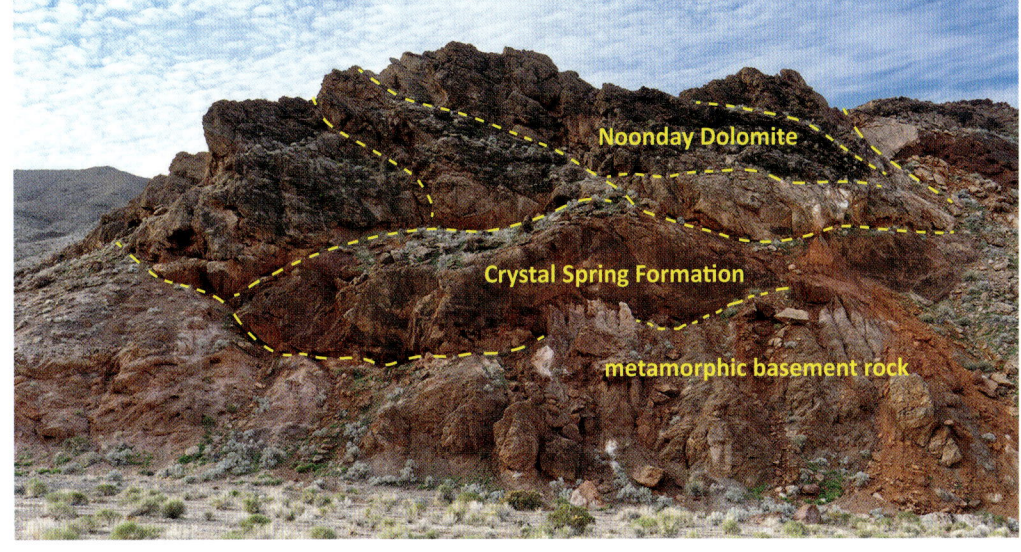

Overall view of Exclamation Point. Nearly every contact—change in rock type—in the photo is a fault.

45

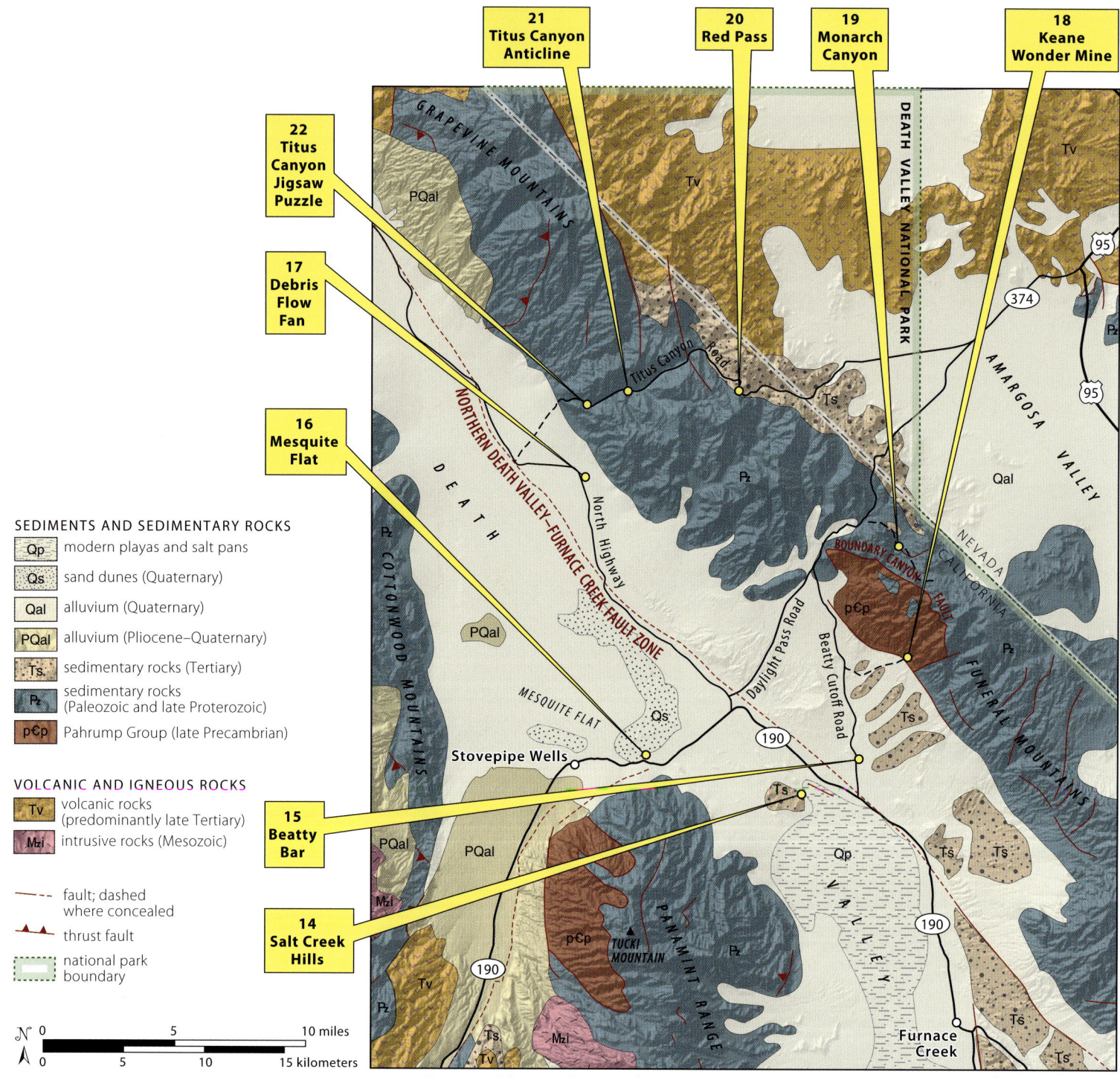

A dust storm over the Mesquite Flat sand dunes obscures the view of the distant Grapevine Mountains.

NORTHERN DEATH VALLEY

Death Valley jogs slightly west where its northern part follows a northwesterly trend along the Northern Death Valley–Furnace Creek fault zone. Tucki Mountain, at the northern end of the Panamint Range, juts into the valley, stopping windblown sand, which piles up at the southern end of Mesquite Flat.

Northern Death Valley lies in the morning shadow of the Grapevine and Funeral Mountains to its east. Although no single mountain range in Death Valley presents the region's entire geologic history, the Funeral and Grapevine Mountains come close. Their oldest sedimentary rocks are part of the Pahrump Group, which records the formation and breakup of the supercontinent Rodinia in Proterozoic time. Above them, another 30,000 feet of sediment preserves evidence of the nearshore marine conditions that prevailed throughout much of the late Proterozoic and Paleozoic. During the Mesozoic Era, these rocks experienced crustal compression, resulting in thrust faulting, folding, and high-temperature metamorphism.

In addition to the most-recent period of crustal extension that began between about 16 and 14 million years ago, these ranges also show evidence of earlier periods of extension. The metamorphic rocks of Monarch Canyon and the Keane Wonder area experienced a second metamorphism at cooler temperatures because they were uplifted partway through the crust during the last part of the Mesozoic Era. The Boundary Canyon fault, which separates non-metamorphosed sedimentary rocks of the Grapevines from the metamorphosed sedimentary rocks of the Funerals, likely started moving during this early time, bringing the metamorphic rocks up from their deep burial. The Titus Canyon Formation, exposed at Red Pass (site 20), speaks to substantial topography that existed between 37 and 30 million years ago, likely the result of extension. Today, the straight, steep range fronts display wineglass canyons that formed because of the recent and ongoing uplift.

One-inch-long pupfish and the sandy bottom of Salt Creek.

SALT CREEK HILLS
Please Don't Pet the Pupfish

The low Salt Creek Hills rise out of the valley floor, uplifted by the same faulting and folding that brings groundwater to the surface. Salt Creek, mostly a groundwater flow path that connects northern Death Valley to Badwater Basin, flows through the hills year-round, supporting life in the middle of the vast desert floor. The life-giving surface water of Salt Creek issues from McLean Spring and Salt Creek Spring, less than 2 miles upstream from the parking lot. The springs are fed by groundwater beneath Mesquite Flat, which flows southward through gravels of the recent alluvial deposits and, to some extent, the underlying Funeral Formation. Beneath the permeable gravels lies relatively impermeable siltstone of the Furnace Creek Formation. At McLean and Salt Creek Springs, faulting and folding of the rock causes the Furnace Creek Formation to rise to the surface, which forces the groundwater upward. The stream, much of which is accessible by boardwalk, has cut a small canyon through the hills. As you walk the half-mile boardwalk, you can inspect the siltstones of the Furnace Creek Formation, deposited between about 6 and 2.5 million years ago.

From the boardwalk, you also can look for Salt Creek pupfish (*Cyprinodon salinus*), which are most active during spring months when the water is relatively high and cool. Like other pupfish species, these tiny fish are unique to their specific locality and can survive in the harsh desert climate. They tolerate water temperatures that dip to near freezing in winter and rise to 112°F in the summer, as well as large swings in the water's salinity, which fluctuates as the water level rises or lowers. The water also attracts birds and salt-tolerant invertebrates and supports salt-tolerant plants such as salt bush and pickleweed.

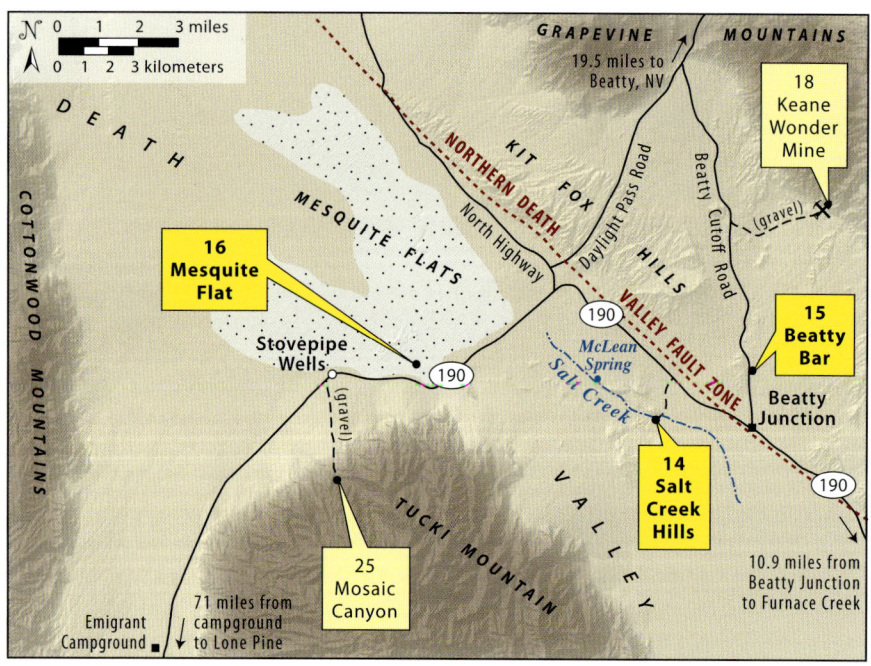

The parking area for Salt Creek is down a gravel road just over 1 mile off CA 190. The Beatty gravel bar is 1.7 miles up Beatty Cutoff Road from CA 190.

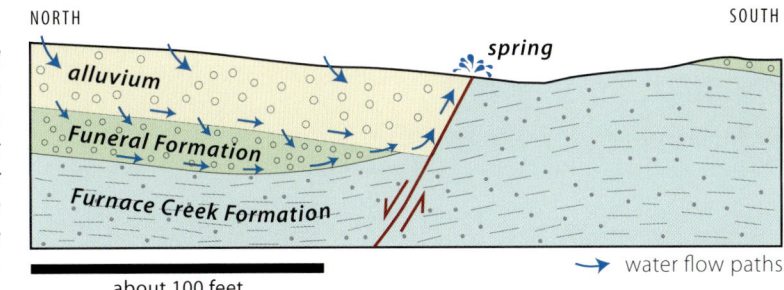

Schematic cross-section illustrating how faulting and folding of the Furnace Creek and Funeral Formations result in groundwater rising to the surface at McLean and Salt Creek Springs. The groundwater flows most easily through the permeable alluvium and, to a lesser extent, the Funeral Formation, but it fails to penetrate the impermeable Furnace Creek Formation. When the groundwater reaches the fault, it rises to the surface.

Salt Creek, lined with salt-tolerant pickleweed, flows through the Salt Creek Hills, which are composed of siltstone of the Furnace Creek Formation.

Looking southwest toward the low Salt Creek Hills in the foreground from CA 190.

15 BEATTY BAR
Gravel Spit in Lake Manly

Imagine standing on a gravel beach with cool wind blowing off a lake that filled Death Valley with hundreds of feet of water. Just 1.8 miles up Beatty Cutoff Road from CA 190 at Beatty Junction, you pass through a low gravel ridge deposited near the shore of Lake Manly, the body of water that existed during the wetter and cooler Pleistocene Epoch.

Based on lakeshore outcrops about 3 miles to the southeast, researchers estimate the highest stand of Lake Manly was 236 feet above sea level. The gravel deposit along the Beatty Cutoff Road, often called the Beatty Bar, lies just below 151 feet in elevation, meaning the lake's shoreline once reached nearly 1 mile farther up the highway. This elevation also suggests the hill to the west was an island. As the climate warmed, the lake receded and likely stabilized long enough at this elevation for strong, wind-driven waves to move the gravel out from the island and deposit it as a spit, a linear beach that juts out into the water.

Looking downslope toward Death Valley, you can see a more subdued gravel ridge about 500 feet away. Along this stretch of highway, a total of five gravel ridges—all spits—have been identified, one of which lies less than 300 feet up the road from here. You can also find fine-grained sediments that were deposited on the eastern edge of the large ridge, indicating water ponded behind the spit. Most researchers agree the ridges formed during the second-to-last high stand of Lake Manly, from 186,000 to about 120,000 years ago.

The roadcut exposes a cross-section of the spit. Look closely to see how much of the gravelly pieces are rounded and in contact with each other, unlike alluvial fan gravels that are typically angular and encased in silt and clay. Given the cobble sizes of some of the gravels, one must wonder at the wave power—and the wind necessary to cause the waves. Jeff Knott and his students at Cal State Fullerton estimated wind velocities of 31 to 60 miles per hour produced the waves necessary to move the cobbles to their present locations. These wind velocities happen on occasion today in Death Valley.

Geologists enjoying the beach at a roadcut through the largest spit. The bedrock hill immediately to the west was an island in Lake Manly.

Crude layering of wave-rounded cobbles deposited in the largest spit.

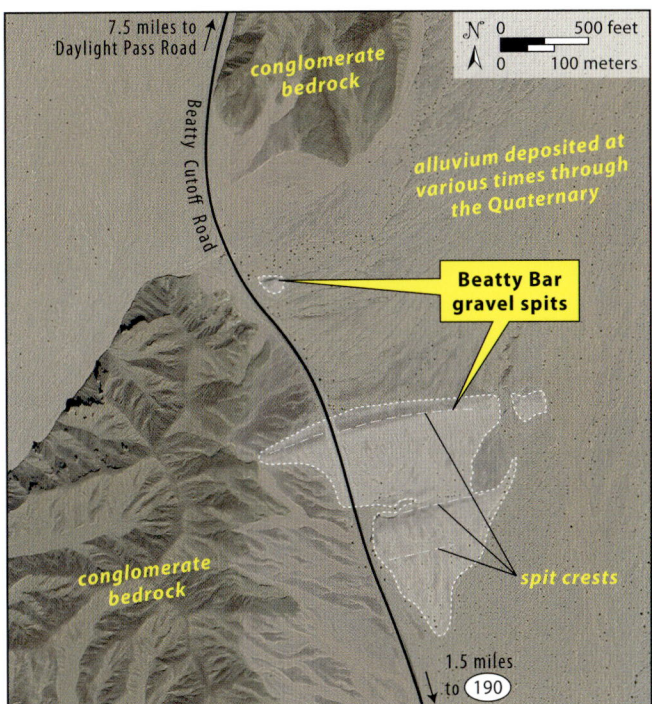

See map on page 48 for a larger overview. This map shows the Lake Manly gravels (gray) and spit crests, hills of conglomerate bedrock of either the Furnace Creek or Funeral Formations, and the lower areas of alluvial fan gravels. Some of the alluvial fan gravels, probably older than 100,000 years, overlap some the Lake Manly gravels, indicating the ridge formed sometime before 100,000 years ago. —Map background from Google Image 2023

View from the "island" looking southeast over Beatty Bar, the ridge with the large roadcut, and two smaller, lower ridges at far right.

MESQUITE FLAT
Star-Studded Sand Dunes

The most accessible of Death Valley National Park's five dune fields, those of Mesquite Flat are perhaps the most interesting. The Mesquite Flat sand dunes offer row upon row of transverse dunes that lead northward to higher star dunes. Mesquite and creosote bush adorn many of the dunes, and vast arrays of wind-formed ripple marks are just about everywhere. Mud-cracked, clay-rich deposits occupy many of the low areas between the dunes and provide short stretches of easy walking into the heart of the field. Instead of singularly tall, steep dunes like Eureka Dunes (site 29) that tend to channel hikers along the same route, these dunes allow people to follow their own paths and find places to reflect on the landscape in solitude.

The dune field collectively encompasses an area of nearly 8 square miles. The popular central part, accessed from a large parking lot along CA 190, lies at the toe of the alluvial fan that spills out of Grotto Canyon. To the south, Tucki Mountain protects the dunes from the prevailing winds that stream up and down Death Valley, allowing windblown sand to accumulate.

As you approach the dunes from the parking area, you walk over the lowest of the fan gravels, composed mostly of limestone and dolomite eroded from Tucki Mountain. Just beyond, numerous creosote bush and mesquite trees growing on the edge of the dune field attest to the presence of groundwater, but these plants gradually thin out and mostly disappear as you reach the center of the dune field. The roots of these plants exert a profound stabilizing effect on the sand. Wind-eroded trenches that partially encircle the plants demonstrate their ability to survive despite the harsh environment.

Before reaching the higher, star-shaped dunes, you'll cross numerous lower dunes that are elongate in a roughly northeast-southwest direction. Called transverse dunes, these dunes form perpendicular to the prevailing wind direction and migrate in the direction of their steeper sides; wind blows sand up their gentle, southeastern sides to where it reaches the dune crest and then avalanches down the steeper, northwestern sides. Notice the multitude of ripples that adorn the gentle sides of the dunes and tend to run parallel to their crests, mimicking the process in miniature.

Between many of the transverse dunes, polygonal mud cracks decorate the fine-grained, clay-rich sediments. These sediments were deposited at various times when the low areas between dunes were partially inundated with water that flowed off the

View north down an arm of a star dune, with wind ripples in the foreground.

Grotto Canyon fan to the south or Mesquite Flat to the northwest. It's tempting to think these sediments reflect a preexisting lakebed over which the dunes formed, but you'll notice they exist at varying elevations in the dune field and clearly overlie—and underlie—much of the sand.

The star dunes, the tallest dunes at Mesquite Flat, rise abruptly above the transverse dunes near the north edge of the field. Their three or more arms radiate from a central peak to make a star-like shape when viewed from above. The varying orientations of the arms indicate inconsistent wind directions.

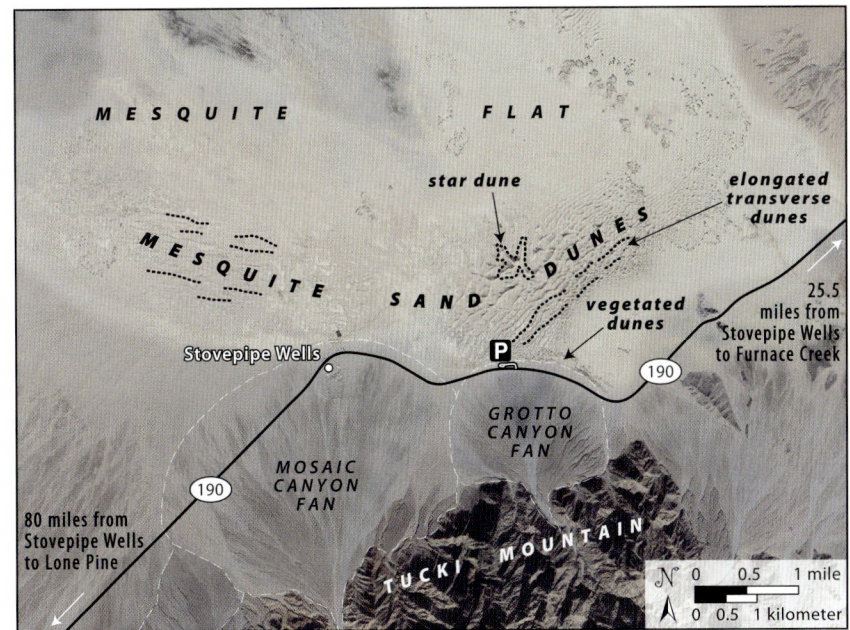

The Mesquite Flat dune field is accessed from CA 190, just under 2 miles east of Stovepipe Wells. See map on page 48 for a larger overview. —Map background from Google Image 2023

Interdune areas occasionally flood with water during large storm events, depositing fine-grained sediments that crack when they dry.

The fan is best viewed from a little way to its north.

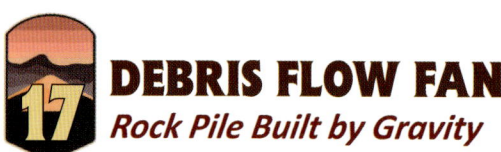
DEBRIS FLOW FAN
Rock Pile Built by Gravity

On the North Highway, just under 12 miles north of the intersection with CA 190, is a pullout at the base of an unusually steep fan. It is best viewed from a short distance north of the pullout, beyond milepost 12 as the highway curves to the west. The steep cone of loose rock is a debris flow fan, formed by repeated debris flows that start in the Grapevine Mountains behind it. During or after storms, a fluidized mass of variable-sized rock is carried rapidly downward in response to gravity. In contrast, transport of alluvial fan sediment occurs primarily through water but is aided by gravity. The large aprons of sediment at the base of the Cottonwood Mountains, across Death Valley to the west, are alluvial fans.

You can see multiple debris flows on the fan, distinguished from one another by varying shades of gray or brown. Over time, an exposed rock in the desert becomes darker due to the accumulation of manganese and iron oxides. Known as desert varnish, the coating becomes fixed to the rock because of the combined actions of water, windblown dust, and biological processes. The older debris flows are shaded more darkly than the younger, light-gray ones. This fan shows at least four different shades of varnish on its surface, indicating at least four different flows, but there is an untold number of flows beneath them.

A close look at one of the newest, light-gray flows reveals its surface consists of a pile of loose, angular rocks that range in size from cobbles to boulders. In some places, you can peer deeper into the flow to see finer-grained material. During active flowing, this lower zone is saturated with water and fluidized, carrying the load of mixed rock on its back. As it cascades rapidly down the slope, larger rocks tend to rise through the mass toward the top of the flow.

View of the debris flow fan from the kiosk at the pullout. It doesn't show the steepness of the fan, however, as well as from other angles.

Individual debris flows show varying shades of gray and brown depending on the amount of desert varnish that has accumulated through time. This light-gray flow is one of the youngest.

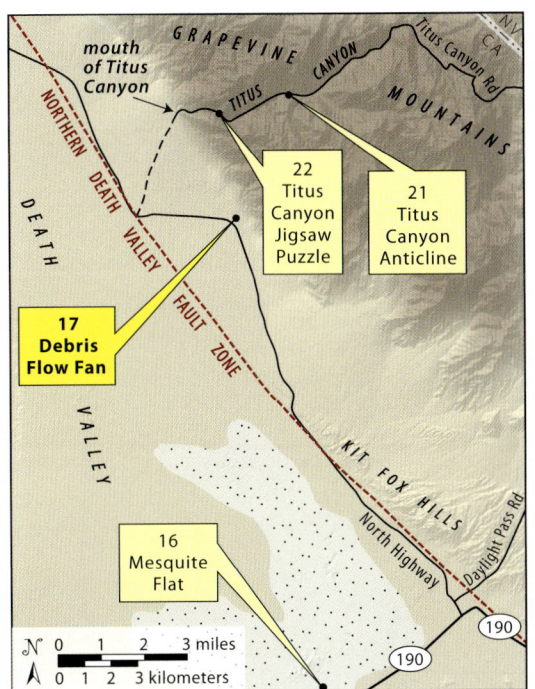

A debris flow fan almost reaches the North Highway at the highway's closest approach to the Grapevine Mountains.

18 KEANE WONDER MINE
There's Gold in Them Thar Metamorphic Rocks

You don't have to hike the 1.3-mile trail up 1,500 vertical feet to experience the Keane Wonder Mine, one of the few productive gold mines in the region. The stamp mill at the bottom of the trail is awesome, and the aerial tramway that's strung across deep canyons is positively mind-boggling. During its heyday, from 1907 to 1912, buckets carrying 70 tons of gold-bearing ore apiece traversed the tramway's cables.

The rocks of the Keane Wonder area from the parking lot to the mine all originated as sedimentary and igneous rocks of the Crystal Spring Formation, the lowest part of the Pahrump Group. During the Mesozoic Era, the sandstone, shale, limestone, dolomite, and diabase sills were buried and metamorphosed. The sandstone was transformed to quartzite by the heat and pressure, the shale became schist, the limestone and dolomite became marble, and the diabase became amphibolite. Colors help identify these rocks: the schists tend to be greenish gray, the marbles light tan to brown, and the amphibolite dark gray to black. Researchers who sampled near the crest of the range determined the highest temperature of

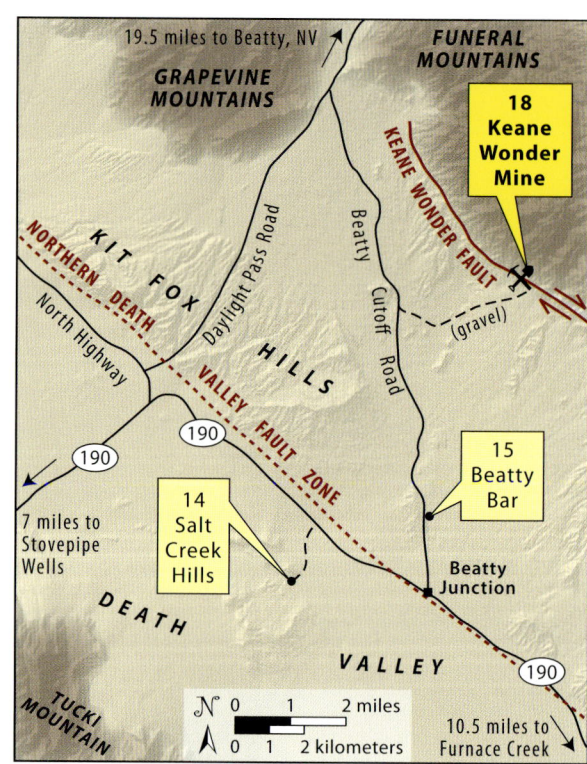

About 5.7 miles north of Beatty Junction, turn east off Beatty Cutoff Road and drive 2.8 miles on a gravel road.

At the Keane Wonder stamp mill, at the bottom of the trail, the ore was crushed to extract the gold.

metamorphism was about 575°C and occurred at a depth of about 15 miles between 167 and 165 million years ago. If you hike the trail to the mine, you'll surely find schists with little, round, red-brown crystals of garnet and gray, blocky crystals of staurolite.

The green coloration in the schists comes from the mineral chlorite, which formed as the rocks were further metamorphosed, but at lower temperatures, between 153 and 146 million years ago. The area that exhibits this later, low-temperature metamorphism coincides with much of the gold mineralization, suggesting the two might be related. The right-lateral Keane Wonder fault separates this part of the Funeral Mountains from the Furnace Creek Formation and some older sedimentary rocks that form the low hills poking through the alluvial fan. Travertine deposits, calcium carbonate deposits precipitated out of spring water, have accumulated along a fault's trace just northwest of the parking area at Keane Wonder Springs.

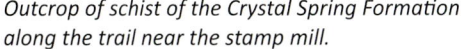

Outcrop of schist of the Crystal Spring Formation along the trail near the stamp mill.

The trail to the mill climbs through marble (brown and tan) and amphibolite (dark greenish gray) along the route of the aerial tramway. The rounded peak at the skyline is capped by the Proterozoic-age Johnnie Formation.

MONARCH CANYON 19
Metamorphic Mayhem in the Funeral Mountains

There may be no better place in Death Valley to see folds, shear zones, and other features that formed due to high-temperature deformation than in Monarch Canyon. Between 167 and 165 million years ago, the rock experienced such high temperatures that some of it melted. The rocks were deformed again during two periods of crustal extension, the oldest of which was probably about 70 million years ago, near the end of the Mesozoic Era, while the most recent took place in the last 10 million years.

The upper 1.4 miles of Monarch Canyon are easily walkable and can be split into three sections: an upper drivable part, and a middle and a lower part. Each contains its own unusual features, making this hike down a relatively secluded canyon one of constant discovery. It ends where the vegetation growing near Monarch Spring renders further passage extremely difficult.

UPPER SECTION
(APPROXIMATELY TWO-THIRDS MILE)

After walking or driving only about 300 feet, you cross the Boundary Canyon detachment fault, which separates unmetamorphosed red sandstone and tan carbonate beds of the Stirling Quartzite to the north from the highly metamorphosed rock of the canyon. In descending order south of the fault, you pass through greenish schist of the Johnnie Formation, ledges of the Kingston Peak Formation, and finally marble cliffs of the Beck Springs Dolomite, all Proterozoic in age. Notice the discontinuous bodies of light-colored pegmatite, an igneous rock with unusually large crystals, that thicken and thin in the Kingston Peak Formation and Beck Springs Dolomite. Called boudins, they form because of stretching parallel to their long dimensions. As the canyon narrows between light-tan and brown walls of Beck Spring Dolomite, you'll see countless folds of all sizes and shapes as well as small ductile shear zones that show relative movement but don't actually break the rock.

This drivable, upper section ends at several large pegmatite outcrops above a dry fall. Notice the complexly folded schist and pegmatite just below the dry fall. A side canyon with its own spectacular features joins the main canyon from the east.

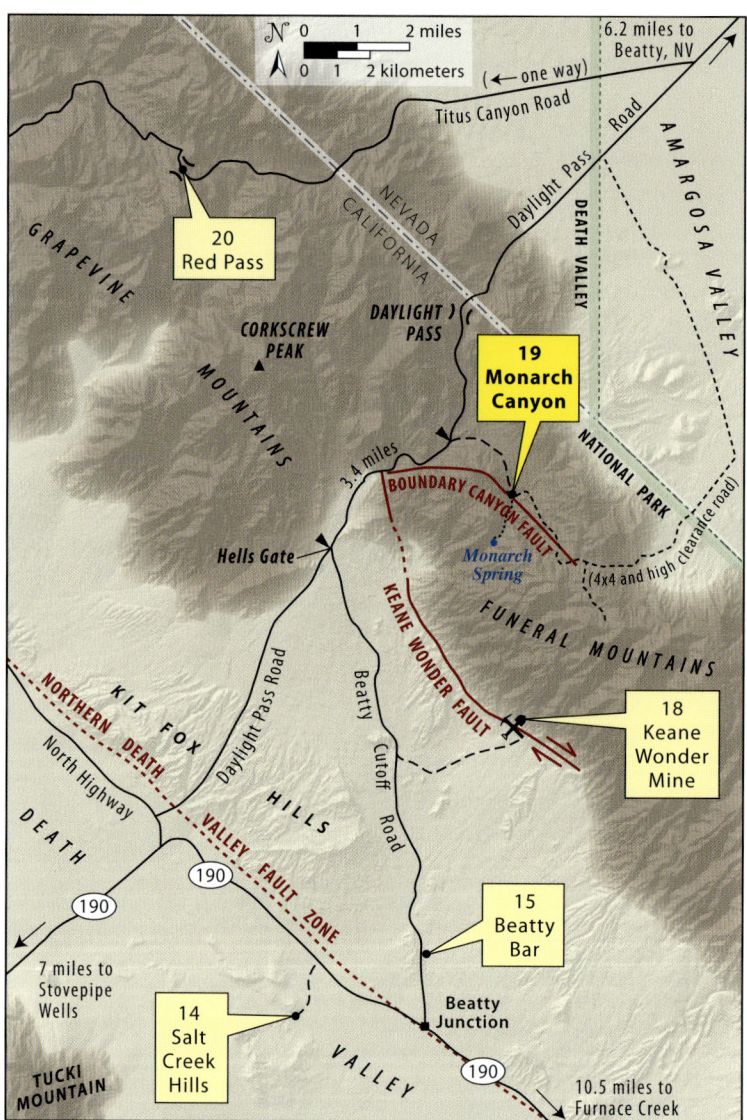

From Daylight Pass Road, turn east just over 3.3 miles north of Hells Gate or 2.8 miles south of Daylight Pass and follow the gravel road (high clearance necessary) 2.2 miles to its hairpin turn. Park the car in a suitable spot or drive a half mile down the canyon and start from the top of the dry fall.

Pegmatite boudins in schist of the Kingston Peak Formation.

View up the upper section of Monarch Canyon past brownish outcrops of metamorphosed Beck Spring Dolomite to the Boundary Canyon fault, marked by the arrow.

Folded Horse Thief Springs Formation in the middle section.

View down the middle section of Monarch Canyon to the Death Valley floor and Tucki Mountain.

MIDDLE SECTION
(APPROXIMATELY ONE-HALF MILE)

A well-maintained trail bypasses the dry fall on its south side, dropping into schist of the Horse Thief Springs Formation. At the canyon bottom, you encounter hardened debris flow deposits, much like the polished ones of Mosaic Canyon (site 25). In the schist of the west canyon wall, you can find elongate, light-gray crystals of kyanite. Researchers use its presence, along with other minerals in the rock, to suggest these rocks were metamorphosed at depths greater than 15 miles.

As the canyon bends gently westward, schist and marble are deformed into a large U-shaped fold on the northwest wall. Some especially good examples of the kyanite appear in the schist just past the bottom of the fold. Almost directly across from the abandoned gold mine workings, you can find some quartzite that was deformed at high temperatures in a ductile shear zone. The rock is very finely crystalline and platy, with its thin layers marked by a distinct linear pattern that trends roughly east-west, parallel to the direction of shearing.

LOWER SECTION
(APPROXIMATELY ONE-QUARTER MILE)

The lower part of the hike starts where the canyon takes a 90-degree turn to the northwest and cuts through a ductile shear zone that some researchers call the Monarch Canyon fault. The canyon wall at this point consists of gneiss with numerous asymmetric mineral or rock fragments that indicate northwestward shear.

Below the turn, you see banded gneiss that probably originated as Crystal Spring Formation. The gneiss contains layers of amphibolite as well as numerous granitic bodies that either follow the layering or cut across it. Most researchers agree that the granitic rocks originated when the rock became so hot that it partially melted, producing a metamorphic rock called migmatite.

Just before the canyon bottom becomes choked with vegetation at Monarch Spring, look for an outcrop on the west side marked by large, reddish garnet crystals. Scramble up ledges of beautifully banded gneiss on the same side for a view down the canyon. Some folks might want to continue through the vegetation to an impassable dry fall less than a quarter mile away, but it's a tough go.

The bottom of the lower section becomes choked by vegetation that thrives on water provided by Monarch Spring.

Partially melted (migmatite) gneiss in the lower stretches of the lower section.

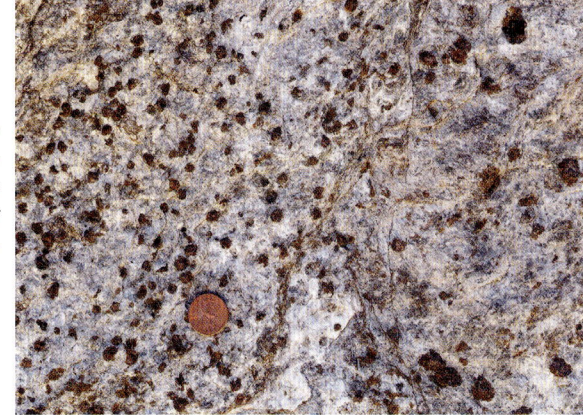

Garnet-rich gneiss in the lower stretches of the lower section.

20 RED PASS
The Colorful Titus Canyon Formation

Red Pass, a divide in the Grapevine Mountains utilized by Titus Canyon Road, is a great place to survey the geologic history of the last 40 million years, as well as the Cambrian Period some 500 million years ago. The pass provides the most-colorful view on Titus Canyon Road, with green, tan, and gray rocks in addition to the bright-red siltstone and sandstone of the Titus Canyon Formation, which outcrops at the small parking area. The Titus Canyon Formation consists of three parts: a base composed of brecciated limestone and dolomite, a middle section of red siltstone and sandstone with scattered conglomerate beds likely deposited as debris flows, and an upper multicolored part of mixed sedimentary rock, which includes an abundance of conglomerate deposited in river channels. You drove along outcrops of the basal part to reach the pass and will descend through much of the upper part as you reach the abandoned town of Leadfield about 3 miles down the valley. The middle part lies just east of the road at the pass.

Rocks of the Titus Canyon Formation form a narrow belt that extends southeastward from the upper reaches of Titus Canyon to the east side of the Funeral Mountains, some 22 miles away, including the low, rounded hills along the gravel road to Monarch Canyon (site 19). Its age, reported in 2022 and 2023 to be between 37 and 30 million years, points to a little-known period of extensional mountain building. The breccias of its base formed where rocks tumbled down steep topography as talus or debris flows, while the two other units were deposited in various high-gradient river environments. Studies of the gravel content of the conglomerates near the base suggest they came from a western source, likely the rocks now exposed in the Cottonwood Mountains. Younger rocks in the sequence, however, probably came from a source in northern Nevada.

The sedimentary rocks above the Titus Canyon Formation consist of the green, mostly conglomeratic Panuga Formation and overlying tan-colored Wahguyhe Formation. The Panuga has an age of 15.7 million years and was deposited after an erosional period that prevailed after deposition of the Titus Canyon Formation. From the red beds at the pass, you can walk northeastward

View northwest from Red Pass down valley eroded in mostly the upper part of the red and green Titus Canyon Formation (TC). Overlying rocks include the Panuga Formation (P), which displays a similar green color as parts of the upper Titus Canyon Formation, tan-colored Wahguyhe Formation (W), and reddish volcanic rocks from the Timber Mountain caldera (V).

across a fault into cliffs of the Panuga Formation and then scramble up a 700-foot peak, the upper 200 feet of which is composed of ash-rich sandstone and altered tuff of the Wahguyhe Formation. Looking northwest toward Leadfield, you can see even younger red and brown layers of volcanic rock framing in the valley. They were erupted from the Timber Mountain caldera, some 30 miles away in Nevada, about 11.5 million years ago.

It's all downhill from Red Pass to the mouth of Titus Canyon. Except for a brief excursion into the Bonanza King Formation in some 600 feet as you approach the first sharp turn in the road, you'll drive through the upper part of the Titus Canyon Formation for the next 3.5 miles. Just past the abandoned mining town of Leadfield, the road turns sharply left into the Bonanza King Formation, a carbonate unit deposited in quiet, warm seawater in Cambrian time.

At Red Pass, the red siltstone and sandstone at the bottom are red beds of the middle Titus Canyon Formation, the cliffs of conglomerate above that are part of the Panuga Formation, and the light-tan rocks above that are the Wahguyhe Formation.

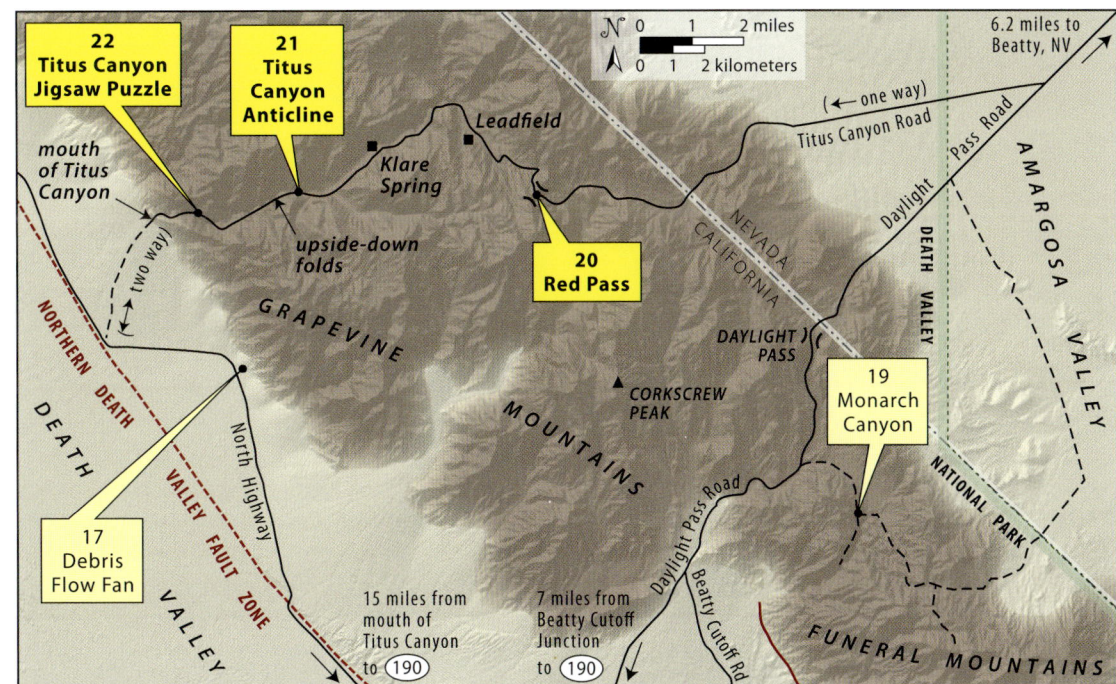

You'll need high clearance and at least a half day to drive the one-way Titus Canyon Road, which you access from NV 374 just over 6 miles west of Beatty. Site 20 is about 12.5 miles, Site 21 is about 19.5 miles, and site 22 is just over 23 miles along Titus Canyon Road.

TITUS CANYON ANTICLINE
Overturned Fold in the Zabriskie Quartzite
See map on page 63.

Titus Canyon Road features one jaw-dropping sight after another, but for geologists, a single outcrop exemplifies its amazing geology: a steeply inclined rib of reddish-purple quartz sandstone that traces out a mountain-scale fold that affects all the rocks from there to the canyon mouth. Look for the outcrop a few steps off the north side of the road and 1.4 miles down canyon from Klare Spring and the rock art petroglyphs. The sandstone is part of the Zabriskie Quartzite, deposited during the Cambrian Period when the Death Valley area oscillated between a shallow ocean and a coastal environment.

Coastal rivers deposited the sand in a rippled form (as small dunes), which appears in cross-sectional views as cross-bedding—thin, curved layers inclined to the overall bedding in the rock. If you look at it closely, you can identify the top of each bed because it's cut off by the overlying bed. The cross-beds curve into parallelism with the underlying bed on which they were deposited. These rocks get younger toward the west, or down canyon.

Ordinarily, rocks are inclined in the direction they get younger, but here they dip eastward because they're overturned. Indeed, you just drove through the older Wood Canyon Formation up canyon. If you take a few steps back and follow the Zabriskie rib upward to the skyline, you can see it folds over to a near horizontal position. You're near the center of the Titus Canyon anticline, an overturned fold formed during Mesozoic crustal compression.

From here to the mouth of Titus Canyon, the rocks are overturned—and in some places, they look nearly horizontal because they are completely upside down. In only a quarter mile, you'll pass through the younger Carrara Formation and from there into the gray dolomites of the Bonanza King Formation. In 0.7 mile, you'll see two large folds in the south canyon wall that are also upside down.

Aerial photo of Titus Canyon anticline showing the Wood Canyon Formation (W), Zabriskie Quartzite (Z), Carrara Formation (C), and Bonanza King Formation (B). The rocks in the canyon (base of photo) get younger toward the west.

The holy grail of Titus Canyon outcrops displays cross-bedding in the Zabriskie Quartzite, outlined for clarity in the red box. The cross-bedding indicates the rocks are overturned. The yellow arrow points to the "up" direction in the rocks.

Overturned Zabriskie Quartzite next to Titus Canyon Road. The quartzite continues up to the top of the ridge where it turns back over to right-side up and is labeled Z; the older Wood Canyon Formation is labeled W.

TITUS CANYON JIGSAW PUZZLE
Breccia in the Bonanza King
See map on page 63.

Just over 3.5 miles below the Zabriskie Quartzite site and less than 1 mile from the canyon mouth, the flood-polished walls of Titus Canyon are broken into a wonderful mosaic of angular gray dolomite blocks of the Bonanza King Formation swimming in a matrix of white calcite. Called a breccia, the jigsaw puzzle fit of many of the blocks has puzzled geologists for a long time. Breccias, rocks composed of large angular fragments, form in a variety of ways. They can form in fault zones, where rocks are broken and recemented; they can form near cliff bottoms or caves, where broken angular rocks accumulate and later get cemented together; or they can form at the base of lava flows, where cooling lava solidifies but then gets ripped apart as the overlying lava continues to move.

Given the Bonanza King Formation is sedimentary and there are no throughgoing faults nearby, we can eliminate the fault zone and lava flow origins. The broken pieces of the rocks could have accumulated as talus from rock falls, but how did so many of the blocks become suspended in the calcite matrix? You would expect them to be resting on top of each other, with calcite simply filling the intervening spaces. A possible answer comes from studies of calcite- and quartz-filled veins in rocks. Many of these veins contain multiple generations of crystals, which, along with high fluid pressures, can literally pry the rock apart as they crystallize. It seems plausible that given the right circumstances, such as in a cave with fallen blocks, calcite in groundwater could precipitate over long periods of time, growing between the fallen blocks and eventually, isolating them in the matrix.

You might notice some beautiful crystals of calcite in the matrix, but please remember that collecting specimens defaces the outcrop and is against the law in a national park.

The breccia consists of scattered, angular blocks of gray dolomite surrounded by calcite.

Close-up view of some of the crystalline calcite and broken gray dolomite.

The breccia forms the walls on both sides of the canyon and shows plenty of evidence for scouring by floodwaters that reached 10 to 20 feet above the canyon floor. Note the shadowed alcove in the canyon bottom, also eroded by the floodwaters.

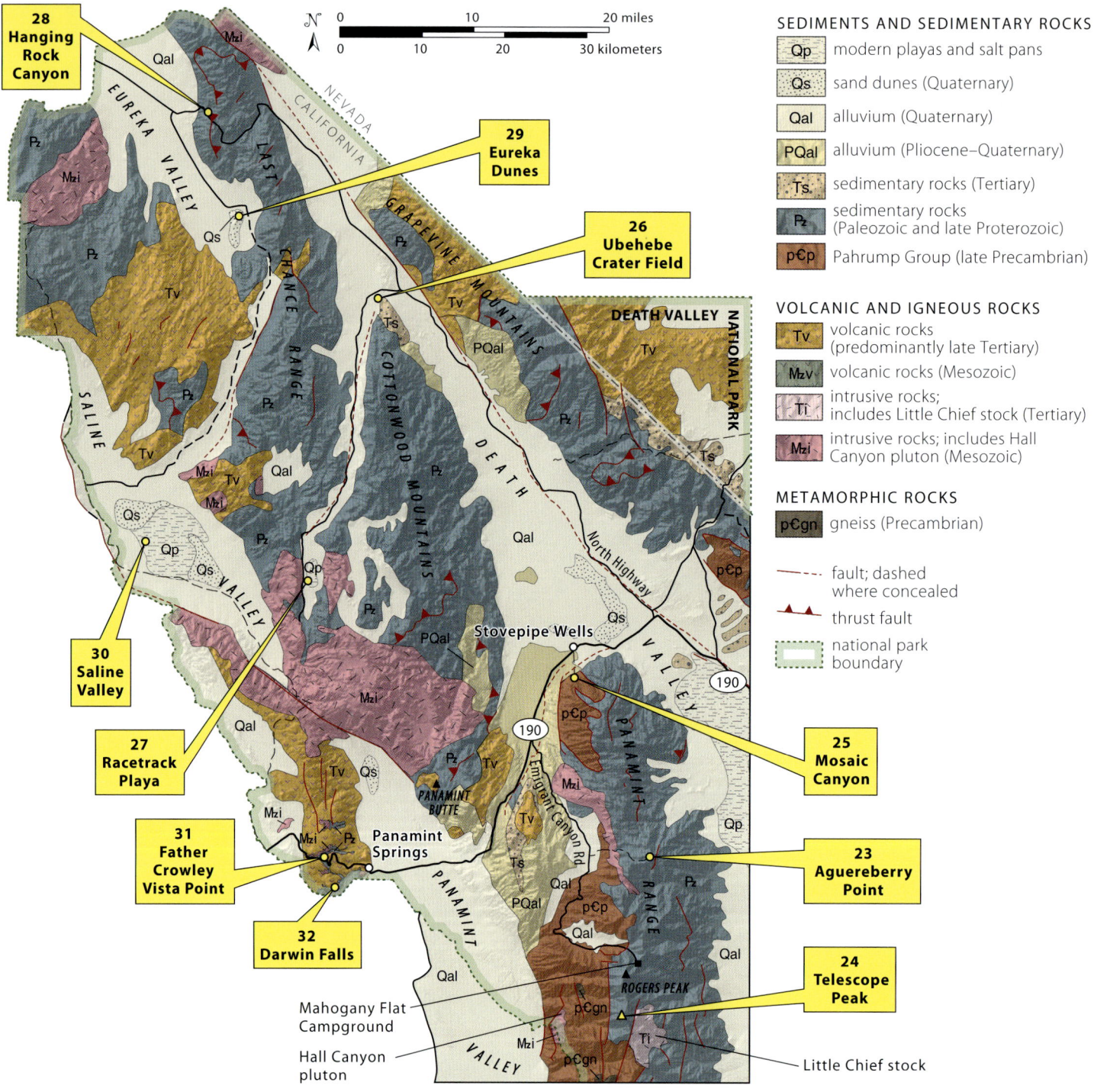

Playa and salt deposits in Saline Valley.

WESTERN RANGES AND BASINS

The western part of Death Valley National Park includes the Panamint Range, Cottonwood Mountains, and Last Chance Range, as well as Eureka Valley, Saline Valley, northern Death Valley, and the east side of Panamint Valley. It's a huge area, more than half the size of the entire park and, except for places along CA 190 and the road to Ubehebe Crater, requires driving on gravel roads to access. These more out-of-the-way sites, however, are certainly worth the effort because they provide insight into both the landscape and the region's geologic history in settings with a greater sense of solitude than the busier parts of the park.

Like the Funeral and Grapevine Mountains, the ranges of western Death Valley National Park display great thicknesses of late Proterozoic and Paleozoic sedimentary rock that experienced folding and thrust faulting during late Paleozoic and Mesozoic crustal compression. However, the western ranges differ in that they were intruded by granitic magmas at different times during the Mesozoic Era and even during the late Cenozoic. The largest intrusion, the Jurassic-age Hunter Mountain batholith, dominates much of the landscape near Racetrack Playa and the southern Cottonwood Mountains. The magmas drove hydrothermal activity that formed gold deposits, responsible for the mining districts like Harrisburg and Skidoo. During crustal extension, basaltic lava flows erupted over the top of many of the older features. These lavas are cut by normal faults that also formed during modern crustal extension.

The valley floors of the western part of the park contain much of the same variability as Death Valley itself, only in a much more remote setting. Eureka Valley hosts Eureka Dunes, the highest dunes in the state. The dunes reach some 600 feet above the valley floor—about 500 feet higher than the tallest star dune on Mesquite Flat. Saline Valley hosts a salt pan and hot springs, and Racetrack Playa, in a high valley between the Last Chance and Cottonwood Mountains, is home to rocks that slide mysteriously across its surface.

23. AGUEREBERRY POINT
Cambrian Cross-Beds above Death Valley

Overlooking Death Valley and much of the Black Mountains at an elevation of 6,433 feet, Aguereberry Point is the preeminent viewpoint from the crest of the Panamint Range. It was named for Pete Aguereberry, the Basque miner who, among other things, discovered the gold deposit in 1905 that became the Eureka Mine and nearby Harrisburg district. You need a high-clearance vehicle to drive to Aguereberry Point.

The viewpoint also offers an unobstructed view of most of Death Valley's Cambrian rocks, beginning with the Wood Canyon Formation that straddles the Precambrian–Cambrian boundary. Just above it, the light-colored Zabriskie Quartzite forms the rock of the viewpoint and prominent white stripe down the canyon wall below. Cross-bedding in the Zabriskie Quartzite indicates the sand was deposited by flowing water, likely in a coastal setting. Cambrian sandstones like it are found throughout western North America as the sea began to inundate the western edge of the continent. The colorful red and brown layers above it are mostly limestones and siltstones of the Carrara Formation, and the gray striped rocks above those are mostly dolomite of the Bonanza King Formation, both deposited in shallow water as the sea continued to rise. The peaks to the south belong to the Nopah Formation, which tops the Cambrian section. This same series of rocks is seen on the west side of Eagle Mountain (site 35), about 40 miles to the east.

The eastward tilting in the Cambrian rock layers is a natural expression of the crustal extension that created the Death Valley area. Both the Panamint Range and the Black Mountains are tilted fault blocks; they rise along normal faults on their west sides and tilt back to the east. From the viewpoint, you can see how the Black Mountains rise abruptly behind the fault zone. Similarly, the Panamint Range rises abruptly behind a fault zone at the edge of the Panamint Valley.

Notice how the salt pan of Death Valley is crowded up against the Black Mountains fault zone, also an expression of eastward

View to the southeast of Death Valley from Aguereberry Point with the Black Mountains in the distance. Sedimentary rocks of Cambrian age tilt eastward toward the Black Mountains fault zone to exemplify the tilted fault block nature of the Panamint Range. The prominent white stripe of the Zabriskie Quartzite outcrops at the overlook. The Wood Canyon Formation, which straddles the Precambrian–Cambrian boundary, underlies the Zabriskie.

Sediment in the Furnace Creek alluvial fan transitions from mostly gravel in its upper part to finer-grained sand and silt at the approximate location of the dashed line, below which water rises to the surface, enabling plants to grow.

Close-up of cross-bedding in the Zabriskie Quartzite.

A high-clearance vehicle is necessary to drive the 6.5-mile road to Aguereberry Point from the paved Emigrant Canyon Road.

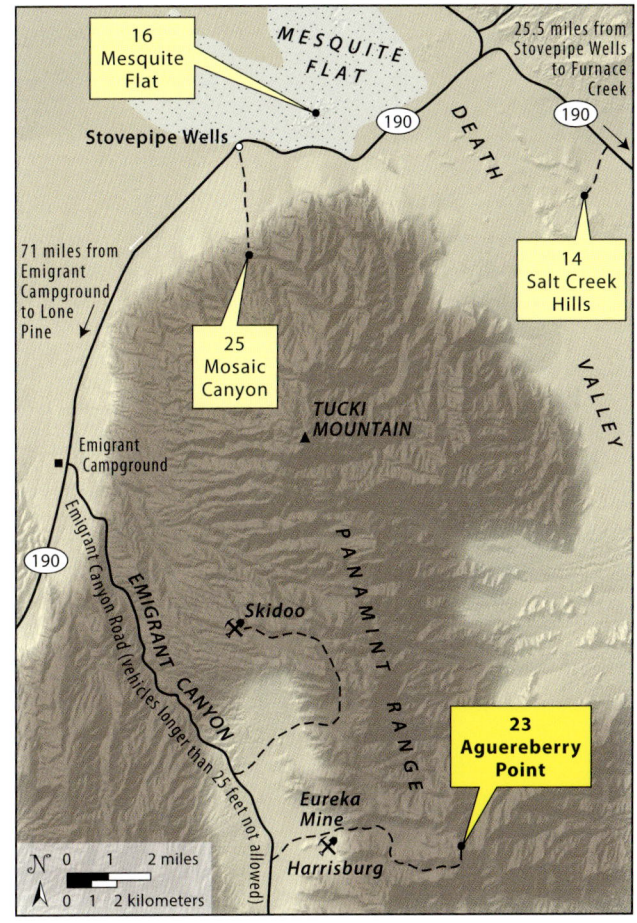

tilting. The salt pan is not tilted very much overall because its surface is continually forming and reforming. Still, the tilting is strong enough to affect the locations where salt can precipitate.

At the north end of the Black Mountains, you can see the Furnace Creek alluvial fan. The dark radial lines are strings of vegetation that grow where springs emerge. Water moves in the subsurface through coarse-grained gravel of the upper fan, and then is forced to the surface when it encounters finer-grained, less-permeable material.

Telescope Peak as seen from 1 mile south of Badwater while the salt pan was flooded in the second half of 2023. The peak is more than 1,000 feet higher than any other point along the ridge. The granitic Little Chief stock forms the light-colored cliffs directly below the peak. Rogers Peak is the next-highest peak along the ridgeline to the north. Mahogany Flat is just off the right edge of the photo.

TELESCOPE PEAK
High Point of a Tilted Fault Block

Reaching an elevation of 11,049 feet, Telescope Peak marks the highest point in Death Valley National Park and is visible from just about anywhere in the park. You don't need to hike the arduous 7 miles to the summit from Mahogany Flat Campground to see the significance of the peak. As a classic tilted fault block, the Panamint Range was uplifted by a fault on its western side, forming a linear mountain range with a long ridgeline. The same rock unit, the late Proterozoic Johnnie Formation, caps almost the entire ridge from Mahogany Flat to 1 mile south of the summit. If you hike the trail, you'll notice most of the rocks are greenish siltstone and some sandstone, with abundant cross-beds, wavy thin laminations, and ripple marks. Only two exposures of the overlying Stirling Quartzite are preserved; the largest forms the summit of Rogers Peak on the ridge about 4 miles north of Telescope Peak.

Telescope Peak practically soars over the rest of the ridge, even though the ridge is composed of the same rock. Typically, as a ridge erodes through time, the less-eroded parts would stand as peaks above the more-eroded parts, but the peaks might differ in elevation by a few hundred feet at most. Telescope Peak, on the other

View of Panamint Valley to the northwest from the summit of Telescope Peak. The snow-capped Sierra Nevada forms the distant skyline.

Wavy cross laminations in the Johnnie Formation as seen along the trail up Telescope Peak.

Looking east from the Panamint Valley, Telescope Peak forms the center of the skyline. The granitic Hall Canyon pluton shows up as the large, light-colored body of rock that makes up the bottom half of the photo.

hand, stands more than 1,000 feet higher than Rogers Peak, the next-highest peak at an elevation of 9,994 feet. From Badwater on the Panamint Range's east side, you can see a possible reason for Telescope Peak's prominence. The Little Chief stock, a body of granite, forms the light-colored shoulder just below the peak and extends vertically downward more than 1 mile. The magma intruded the sedimentary rocks about 12 million years ago at very shallow depths, possibly even venting to the surface. It appears the sedimentary rocks bowed upward to accommodate the granite when it intruded. It's also likely that the heat of the intrusion hardened the nearby sedimentary rocks, making them more resistant to weathering and erosion.

From the Panamint Valley on the west side, you can also see the ridge-like nature of the range, with Telescope Peak forming the highest bump. Low on the front of the range is the light-colored Hall Canyon pluton, a smaller granitic body that intruded the Pahrump Group. Above it, you can see the well-layered sedimentary rocks that generally get younger toward the top of the peak. The Hall Canyon pluton's chemistry suggests it formed by the melting of existing crustal rocks. It also cooled at a depth of about 6 miles between 72 and 70 million years ago—much deeper and earlier than the Little Chief stock. Because it is some distance from the summit, it is unlikely to have a played a role in Telescope Peak's present height.

25 MOSAIC CANYON
Angular Debris Plastered on Wild Marble

On entering Mosaic Canyon, you immediately see how it got its name. Although the canyon is cut into marble of the Proterozoic-age Noonday Dolomite, the walls also display intricate natural mosaics of angular rock fragments suspended in a matrix of silt and sand. The mosaics, within a half mile of the parking lot, are debris flow deposits that periodically fill the canyon. They become polished when hyper-charged flash floods cut through them and scour the canyon walls. New debris flows fill in what's been eroded, and the process continues over and over again. If you look closely at the mosaics, you can see many of them consist of several overlapping deposits.

To get an idea of the power and sizes of some of these debris flows, take a look at the left (south) wall only a quarter mile up the narrow canyon where it takes a sharp right turn. The wall is composed of vertically bedded gravel. On closer inspection, you can see the gravel is part of two gigantic blocks that are surrounded by the finer-grained matrix. The gravel was originally deposited at a near-horizontal angle somewhere up canyon, but these large blocks broke loose and traveled down canyon in a debris flow. They came to rest in a vertical position and then were cemented in place by later deposits. For a visual depiction of today's fluctuating canyon bottom, refer to the excellent description of Mosaic Canyon in Allen Glazner and Art Sylvester's *Geology Underfoot in Death Valley and Eastern California*.

The smoothed, brown-and-white marble walls tell their own story. Marble is metamorphosed limestone. The layering, which

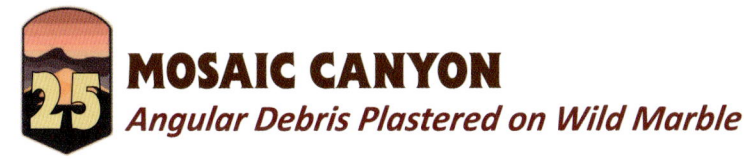

Hiker in narrows of Mosaic Canyon. In this spot, the debris flow mosaics form most of the left-hand side of the photo, whereas the Noonday Dolomite bedrock forms the right-hand side.

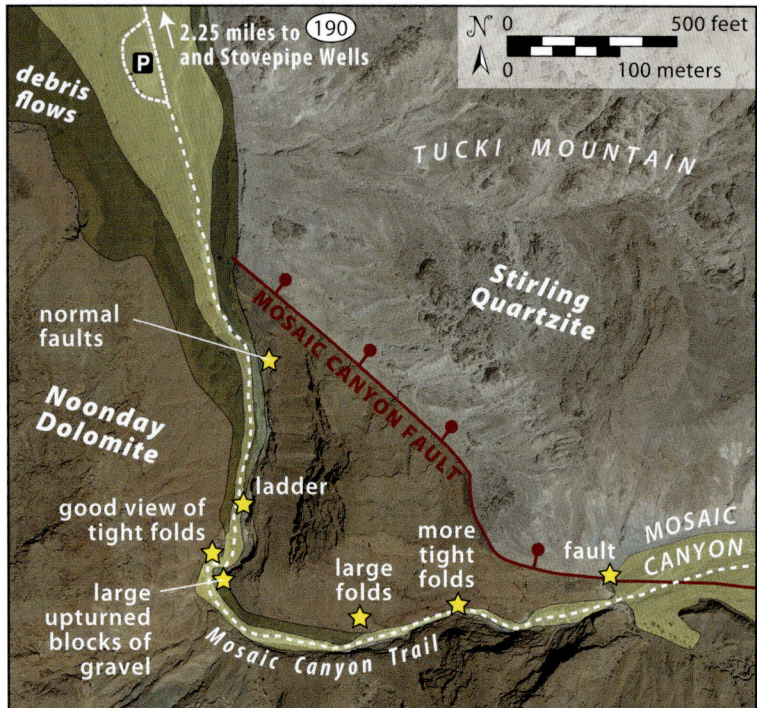

The parking lot and trailhead for Mosaic Canyon lie less than 2.5 miles up a gravel road that intersects CA 190 on the west edge of Stovepipe Wells.
—Google base image, 2013

Overlapping debris flow deposits (top) on bedrock (bottom).

appears nearly horizontal, is not sedimentary bedding but metamorphic layering that formed through the heating, recrystallization, and pervasive folding of the rock when it was hot and malleable. Unlike sedimentary beds that are typically continuous, you can't trace any of the layers exposed in the canyon for more than a few feet. Rather, they pinch, swell, or disappear altogether. Rock fragments contained in the layering appear to have rotated while the material surrounding them flowed past.

Just past the right-hand turn in the canyon, you can see how the rock is folded back and forth over itself. Until this spot, you've been walking southward, parallel to the folds, which makes them difficult to see. Where you walk eastward or westward, however, the folds become dramatically apparent. As you scramble over the slippery bedrock at this point, notice the canyon takes a 180-degree turn. As you walk east, you'll see more and more folds, especially after about 500 feet where the canyon narrows again. Look for two cliff-sized folds, one with a very rounded shape, the other with a severely angular one, just before the canyon narrows. While most researchers agree the folding is related to movement along an extensional fault zone in the hot, middle part of the crust, the timing is uncertain. Some researchers argue it took place during the late Cenozoic, while newer findings suggest it occurred at the very end of the Mesozoic. You can keep walking up Mosaic Canyon another 1.5 miles mile before encountering an impassible dry fall.

Regardless of the timing of the high-temperature folding, you can see several normal faults that break the rocks, all of which slipped during Late Cenozoic extension that formed the Basin and Range. Some of the best examples are exposed in the low eastern cliff just before entering the canyon. They likely relate to slip on the Mosaic Canyon fault, which separates the Noonday Dolomite from the overlying Stirling Quartzite, here a series of greenish and brown sandstones and siltstones. You can see this fault from the parking lot as the edge of the brown Noonday Dolomite east (left) of the canyon mouth. You can also examine the fault's large broken zone where the narrow part of the canyon opens into a much wider area less than a half mile up from its mouth.

Large folds in the Noonday Dolomite.

Blocks of near-vertical gravel on the side of Mosaic Canyon. They are deposited against bedrock of the Noonday Dolomite on the left side of the image.

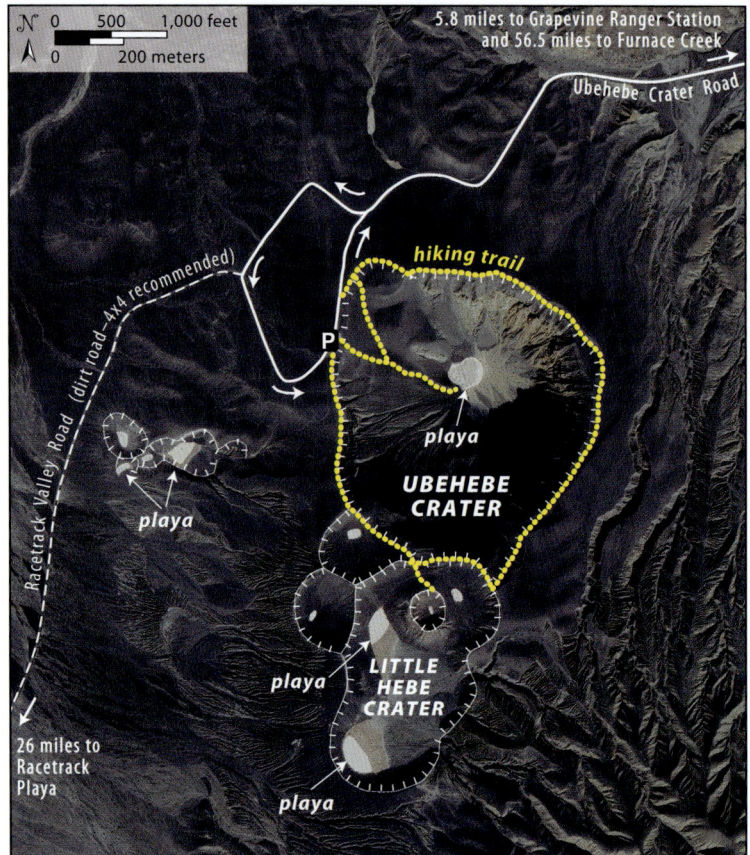

Map of the Ubehebe Crater field. Craters are outlined by white hachured lines, the small playas in their bottoms by light blue, and the hiking trail by the yellow dotted line. Hachures around the crater rims point down toward the crater bottoms. Notice how Ubehebe Crater cuts across the craters to the south. —Google Image, 2023

UBEHEBE CRATER FIELD
Youthful Explosions When Magma Met Water

While so much of Death Valley speaks to its vast geologic history, Ubehebe Crater formed after humans arrived in California. The crater is some 2,500 feet across and 600 feet deep with a nearly vertical eastern wall. It still looks fresh because erosion hasn't had time to soften its appearance. Moreover, some of its ash and cinder deposits, called tephra, overlie archeological artifacts that are likely no older than a few thousand years. Most researchers lean toward 2,100 years as the crater's approximate age. The crater floor is covered by playa deposits from times when it temporarily floods with water.

Ubehebe Crater is the largest of fourteen craters that formed in an area of less than 1 square mile at the northern end of Racetrack Valley Road. The tephra they erupted covers about five times that area. Ubehebe and most of the other craters formed when basaltic magma rising through the underlying alluvial fan deposits encountered groundwater. The resulting explosions likely pulsed with influxes of both new magma and groundwater. The process culminated with the formation of Ubehebe Crater, which partially overlaps the two closest craters and produced deposits that blanket the entire field. You can see the alluvial fan deposits in the crater walls.

The entire sequence of eruptions likely occurred over a short period—perhaps days, weeks, or months, based on a study by Judy Fierstein and Wes Hildreth of the US Geological Survey. Among other things, they found evidence that the youngest Ubehebe tephra was thermally altered where it was deposited on top of deposits of the earlier eruptions, suggesting the older deposits were still hot. The researchers also noted successive deposits don't show any evidence of erosion between them and the chemistry of the erupted rocks changed very little during the eruptive period.

A well-worn trail circumnavigates Ubehebe Crater, leading past several of the older and smaller craters to the south. Each of these smaller craters is also floored by fine-grained sediment from infrequent times of water ponding. Along the trail, you get close-up views of light- and dark-gray and tan Ubehebe tephra, which composes the rim of Ubehebe Crater as well as the tops of the other

Small playa at the bottom of Ubehebe Crater. The color change in the sedimentary rocks of the crater walls marks the location of a nearly vertical fault, which moved before the crater formed.

Panorama looking southwestward at Little Hebe Crater within the three overlapping craters south of Ubehebe Crater.

craters. Its layering, defined by varying grain sizes, reflects variations in the strength of the eruption. The surface of the volcanic field appears dark gray to black because the lighter-colored, fine-grained tephra has largely blown away.

Little Hebe Crater, more than 300 feet wide, occupies a small mound near the center of an elongate depression that consists of three older overlapping craters. These craters mark the earliest eruptions in the field, while Little Hebe formed just afterward. In Little Hebe's walls, you can see a prominent ledge composed of volcanic bombs (large globs of ejected lava that cooled as they flew through the air) and smaller-sized tephra, much of which was welded together from the high temperatures to form the mound. Some of the thermal alteration of Ubehebe tephra suggests only a short time elapsed between successive eruptions. The Ubehebe tephra that directly overlies the ledge is oxidized to a rusty orange.

Ubehebe Crater shortly before sunset. The bright-orange and white sedimentary rocks in the walls of the crater were deposited by streams that flowed into the Furnace Creek Basin.

A sliding rock on Racetrack Playa.

27 RACETRACK PLAYA
Mysterious Movement at Glacial Speeds

Racetrack Playa is home to a remarkable geologic mystery. Set at an elevation of about 3,700 feet in a valley between the southern Last Chance Range and the Cottonwood Mountains, the playa features tracks made by rocks that appear to have slid across the dry lakebed. Since the 1940s, numerous researchers have tried to understand how rocks that weigh anywhere from a few pounds to as much as 700 pounds could slide over the playa surface with no other evidence than an irregular track carved into the playa mud. Many of the tracks exceed 650 feet in length.

The mud becomes very slippery when wet, and the high valley frequently experiences high winds. However, most calculations suggested wind velocities greater than 130 miles per hour are required to move most of the rocks, and those with low profiles still wouldn't move. Furthermore, wind alone doesn't explain why many rocks show identical track trajectories as if they'd moved together. Some early researchers surmised ice was involved. They argued that when the playa froze over during winter months, it could raft together many of the rocks. The ice acted as a large sail, catching the wind and moving the rafted rocks as one.

The answer came in December 2013 when Richard Norris of Scripps Institute of Oceanography and his colleagues used remote photography to catch the rocks moving. At this time, the playa was flooded to a depth of about 4 inches and froze over during the nights. They found as the sun warmed the frozen surface and caused it to break into large, thin panels of ice, light breezes of only 10 miles per hour moved the ice, which then pushed the

Close-up view of the granite of the Grandstand with unusually large crystals of orthoclase feldspar.

rocks. Not all the rocks moved, and the ones that did obtained speeds of only a few inches per second. Through repeated movement events, however, some created tracks more than 500 feet long.

The sliding rocks congregate at the south end of the playa closest to their source, a ridge of Cambrian-age dolomite that is the same age as the Bonanza King Formation. The rocks are folded and exhibit evidence of metamorphism from being intruded by nearby granitic rocks of the Jurassic-age Hunter Mountain batholith. The granitic rocks form most of the mountains on both sides of the playa, including Ubehebe Peak on the west. On the east side of the playa, you can see large, dark bodies of Paleozoic rock surrounded by the light-colored granite.

Near the north end of the playa, a large outcrop of the granitic rock forms the Grandstand, rising some 50 feet above the playa. The rock displays unusually large crystals of orthoclase feldspar. Its unlikely presence within the playa suggests it was part of a landslide that swept into the valley from the north side of Ubehebe Peak, which similarly contains the large crystals.

The 27 miles of gravel road from Ubehebe Crater to Racetrack Playa can be arduous, so inquire about road conditions before making the trip. Also, be sure you have a spare tire and are prepared to spend the night in case you have car trouble.

Scattered blocks of gray dolomite at the south end of Racetrack Playa. Ubehebe Peak is the left of the two conical peaks in the middle skyline.

View of Racetrack Playa and the Grandstand from Ubehebe Peak. The Grandstand rises about 66 feet above the playa.

Looking eastward over a track and rock toward folded Cambrian dolomite (Pz), the main source of the sliding rocks, and the Jurassic granitic rock of the Hunter Mountain batholith (Jg).

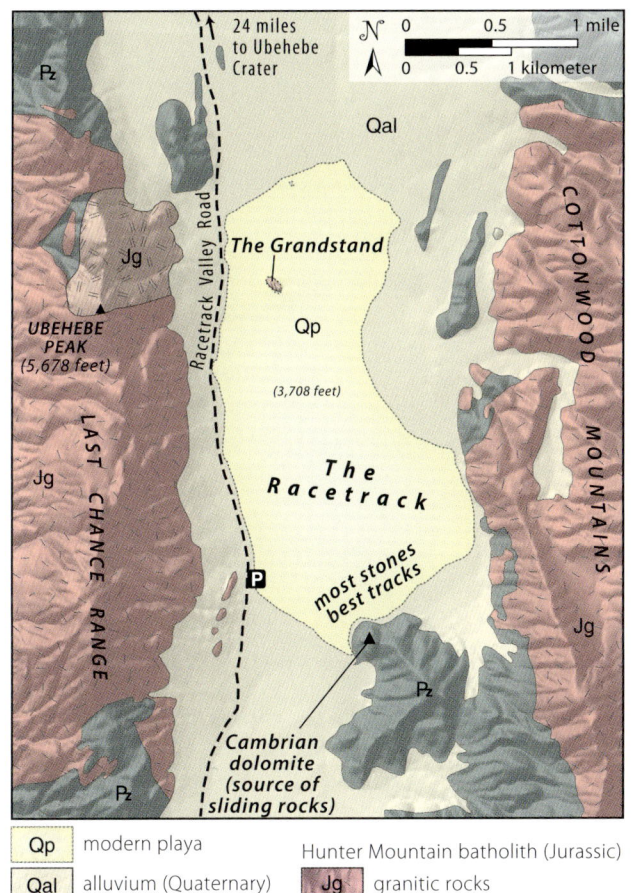

Simplified geologic map of the Racetrack Playa area. The Paleozoic rocks, shown in the green color, consist of a thrust-faulted and folded sequence deposited from the Cambrian through the Pennsylvanian Periods. —Modified from McAllister, 1956

HANGING ROCK CANYON
Drive Beneath Topsy Turvy Thrust Faults

Hanging Rock Canyon, located in the Last Chance Range on the east side of Eureka Valley, presents some of the most complexly folded and faulted rock layers in the entire region. As it descends steeply into Eureka Valley, the canyon cuts straight across the Last Chance thrust system, the oldest and westernmost of Death Valley's thrust faults. Remnants of this regional structure are found disbursed over an area greater than 950 square miles, from the Inyo Mountains near Bishop to the Cottonwood and Grapevine Mountains in Death Valley.

Using geologic relations at Conglomerate Mesa in the southern Inyo Mountains, researchers have inferred the thrust fault moved during the early part of the Permian Period, about 290 million years ago. During this time, the western edge of North America underwent a major change in its plate tectonic configuration, ending its long period of sediment accumulation on the stable continental shelf in generally quiet water. The Last Chance thrust heralded an active tectonic plate margin, characterized by volcanic activity and compressional mountain building.

In Hanging Rock Canyon, you can see intense deformation from the entire series of thrusts that compose the thrust system. Movement on younger fault strands cut off the main Last Chance thrust and overturned it. The main fault is exposed at the west end of the canyon where it places Cambrian-age Zabriskie Quartzite against Mississippian-age Rest Spring Shale. The other thrusts, which are mostly in the folded Cambrian rocks, are visible in the upper half of the north canyon wall. They indicate multiple periods of contraction, including cutting off the original thrust. All of the movement was generally eastward, away from the newly formed plate margin.

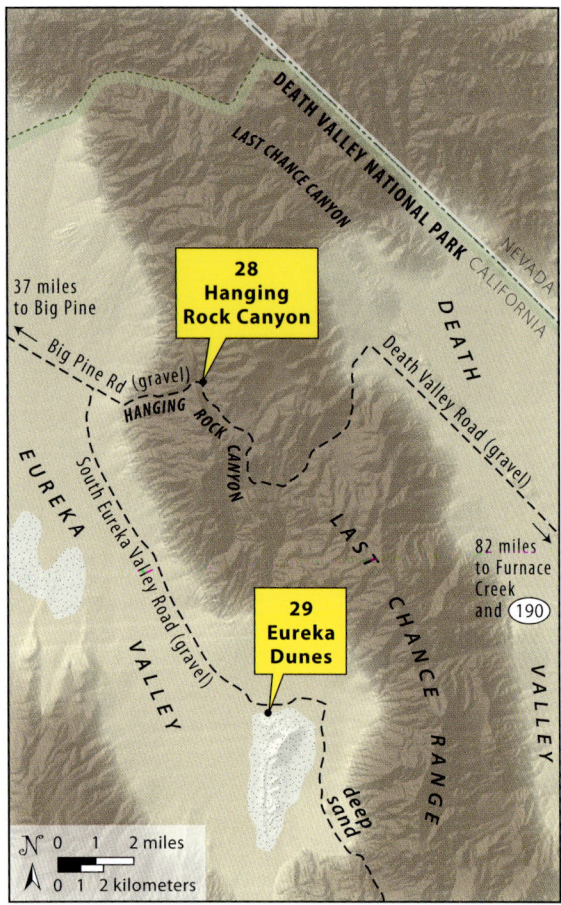

You can reach Hanging Rock Canyon from either side of the Last Chance Range, but the geology described here is on the west side, near its entrance to Eureka Valley.

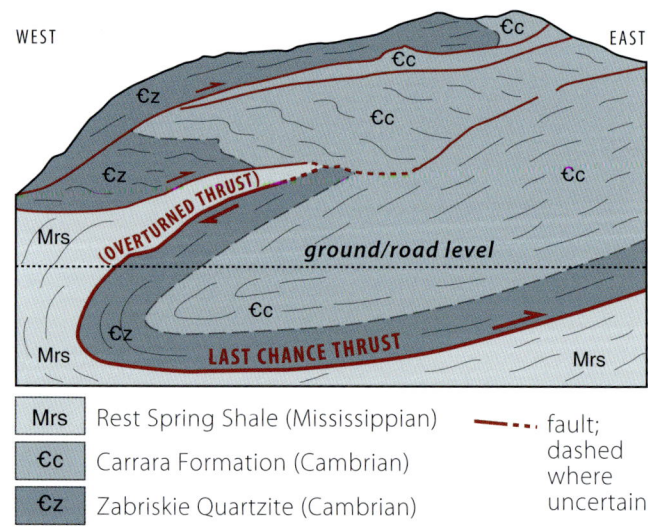

Cross-sectional view of the photo of the north wall of Hanging Rock Canyon (bottom photo on facing page), showing interpretation of Last Chance thrust below ground. Cross section is about 2,000 feet across. —Modified from Levy and others, 2020

The light-colored rock on the skyline, above the entrance to Hanging Rock Canyon, is the Zabriskie Quartzite. Cliffs just left of the road are part of the Mississippian Rest Spring Shale; above them, but not seen in the photo, are outcrops of Carrara Formation and Zabriskie Quartzite.

View of the north wall of Hanging Rock Canyon, showing the locations of the thrust faults (yellow). The Last Chance thrust is overturned.

Eureka Dunes as seen from South Eureka Valley Road. The Last Chance Range behind the dunes consists mostly of Cambrian sedimentary rock.

EUREKA DUNES
Singing the Praise of California's Tallest Sand Pile
See map on page 84.

Rising steeply more than 600 feet off the valley floor, Eureka Dunes are California's tallest dunes. Being in a remote setting, they offer a completely different experience than the dunes at Mesquite Flat (site 16). Eureka Dunes are situated at the southeast margin of the Eureka Valley, where sand carried by winds blowing from the northwest is trapped against the striped Paleozoic rocks of the Last Chance Range. A 10-mile-long, well-graded gravel road connects the dunes to Big Pine Road.

The dune field consists of at least three different types of dunes. Linear dunes, which form approximately parallel to the prevailing wind direction, shape the field into an elongate arc. The highest point of the field lies near the southern end of the largest linear dune, and several lower linear dunes lie to the northwest. Between them are numerous transverse dunes, which are perpendicular to the winds. Several low star dunes, the tallest of which reaches some 250 feet above the valley floor, dominate the field to the south. Deposits of mud and clay cover the valley floor in the area surrounding the dunes. Now broken into countless small polygons, these deposits likely accumulated during wetter periods that caused flooding of the valley floor.

From surveying plant populations, some researchers think the Eureka Dunes are likely the oldest dune field within Death Valley National Park and possibly the Mojave Desert. Compared to other dunes within the park, these dunes host a far-greater variety of sand-loving plants, including subspecies of dune grass and evening primrose (*Swallenia alexandrae* and *Oenothera avita*, respectively) found nowhere else in the world. Because these plants require time to colonize an area, and the endemic varieties require time to evolve, dunes with such diversity must be older than dunes with less plant diversity.

Like most sand dunes, the sand of the Eureka dunes consists of tiny grains of quartz with scattered concentrations of the black mineral magnetite. Unlike nearly all other dunes, however, the Eureka Dunes sing. They are one of fewer than fifty dune fields around the world that can exhibit loud droning sounds when the sand is dislodged down steep, downwind faces. The sounds, which resemble the lower notes from a cello, only occur when dry sand is positioned on steep slopes at least 120 feet above the dune bottom. Studies by Nathalie Vriend of the University of Colorado and her colleagues show the distinctive low frequency sound is produced by constructive interference of sound waves within the moving sand. For the interference to occur, a layer of hard-packed sand must exist underneath the soft surface sand to guide the sound waves. Even though so few dunes exhibit this trait worldwide, it's been reported from other nearby dunes, including Kelso Dunes in Mojave National Preserve, the Dumont Dunes just south of Tecopa, and the Panamint Dunes of Death Valley.

View of Eureka Dunes from their eastern side with mud-cracked playa deposits in the foreground.

Wind ripples in the foreground with the main linear dune in the background. Note how the crests of the smaller transverse dunes run perpendicular to the main dune.

30 SALINE VALLEY
Pure Salt in the Playa

Like Death Valley, diamond-shaped Saline Valley is bordered by high, steep, fault-bounded mountains propelled upward by crustal extension and strike-slip faulting. Alluvial fans along the mountains' edges extend down to a valley floor covered by playa muds, salt, sand, and sand dunes. The Inyo Mountains soar over Saline Valley's southern margin, reaching elevations above 10,000 feet—some 9,000 feet above the valley floor. They consist mostly of Jurassic-age granitic rock that has intruded Paleozoic sedimentary rock.

A normal fault marks the abrupt transition from bedrock to alluvial fan, and numerous features, including fault scarps, wineglass canyons, and triangular facets along the range front, indicate the fault is recently active. Toward the east, the normal fault becomes the Hunter Mountain fault, a strike-slip fault with some 5 to 6 miles of right-lateral slip. The Hunter Mountain fault continues up Grapevine Canyon and joins the Panamint fault to the southeast. With this geometry, continued movement on the Hunter Mountain fault drives normal faulting in both the Saline and Panamint Valleys. Estimates for the slip rate on the Hunter Mountain fault range from 2 to 4 millimeters per year, with the movement beginning to form Saline Valley between 3.1 and 1.8 million years ago.

Overall, the valley's floor tilts gently toward the active normal fault on its west side, which is some 16 feet lower than its east side. And, like Death Valley, the alluvial fans next to the fault are short, steep, and well defined, whereas those on the other side coalesce into a broad bajada, or alluvial apron. The drainages flow toward Salt Lake, the lowest spot in the valley and the most

View to the northeast over the edge of Salt Lake in Saline Valley.

The diamond-shaped Saline Valley is bound by an active normal fault on its southwest side and the strike-slip Hunter Mountain fault zone on its south. Bedrock in the area consists of a sequence of Paleozoic sedimentary rocks intruded by granitic plutons of Jurassic and Cretaceous age, most notably the Jurassic Hunter Mountain batholith (Jg). Salt Lake occupies part of the area of active salt deposition (Qe).

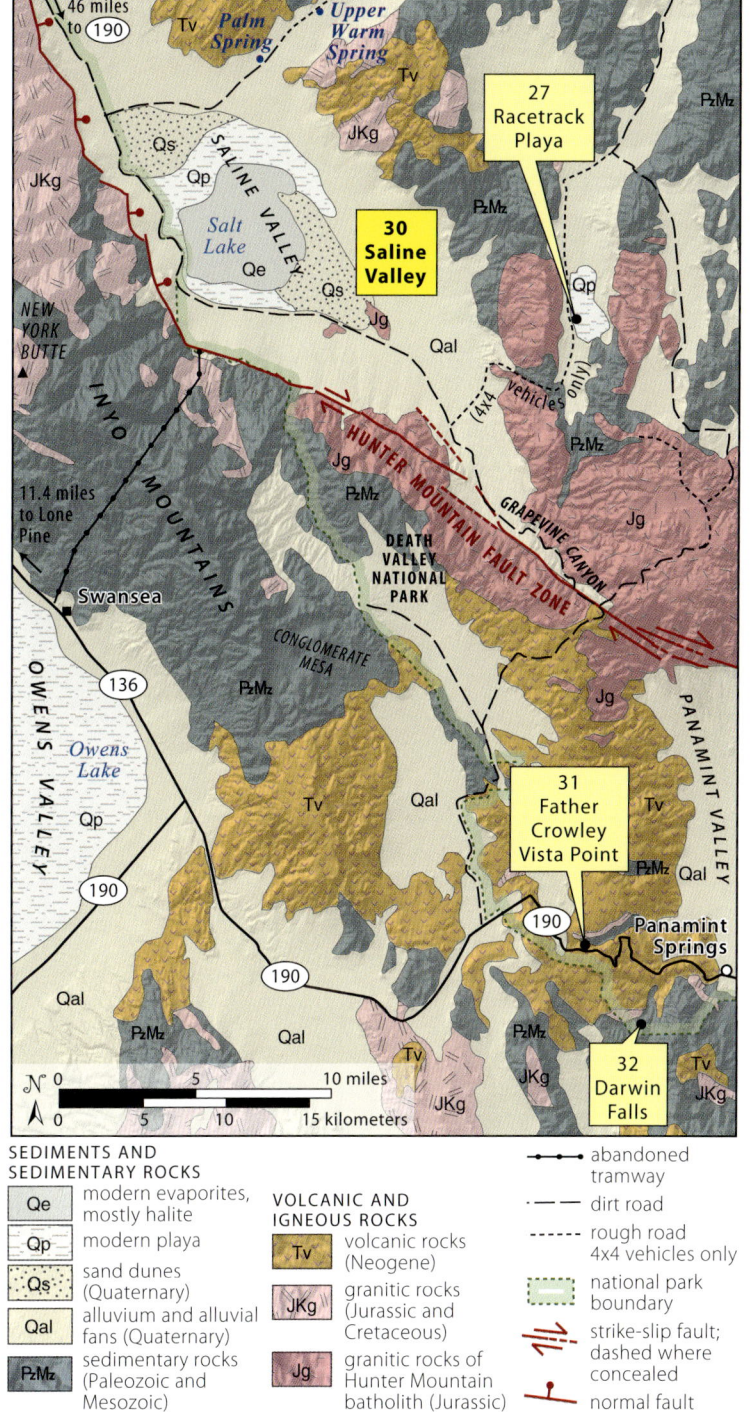

frequently flooded part of the playa, which fills an area just below the fault and steep fans.

The nearly circular playa is largely covered by a variety of salts precipitated from evaporating water. The lake center is mostly halite (sodium chloride), whereas less-soluble sulfate salts such as glauberite and mirabilite occupy slightly higher zones. Gypsum (hydrous calcium sulfate) and various carbonates, being the least soluble, form outer zones around the playa. While some of the water on the playa originates in flash floods that pour out of the canyons, most of it issues from springs. The spring water soaks into the gravel of the alluvial fans and then reemerges on the playa. As the water evaporates, the precipitating salt may contain tiny inclusions of the playa water. Biologists have discovered entire communities of single-celled organisms in these minute water inclusions, and the isolated communities have thrived for tens of thousands of years!

Despite the microbiology, Saline Valley's halite is unusually pure, so much so that it was harvested for a range of uses, including food processing, during the early 1900s. Between 1910 and 1913, the Saline Valley Salt Company constructed a 13.5-mile-long aerial tramway across the Inyo Mountains to Swansea in the Owens Valley in order to expedite shipping to Los Angeles. The tramway carried salt in closed buckets that could hold 700 pounds of salt each. At one point, the tramway transported 100 tons of salt per day. Today, you can still see some of the remnants of the salt works, including salt evaporation ponds and parts of the abandoned tram.

Beyond the playa, the valley floor consists of a large sand sheet that becomes progressively coarser grained toward the

The wineglass canyon cuts through Paleozoic rock at the edge of the Inyo Mountains. The fault lies at the mountain base where the relatively flat alluvial fan meets the steep, whitish rock. New York Butte, which forms the high peak, consists of 134-million-year-old granitic rock and rises to an elevation of 10,668 feet above sea level.

Some remains of the salt works on the valley floor.

surrounding alluvial fans. A small, oval field of sand dunes lies near the northwestern edge of the valley, north of Salt Lake. The field consists of numerous parallel dunes, elongate in a southwest-northeast direction and transverse to the prevailing winds from the northwest.

Three popular clothing-optional hot springs form a north-trending line up a fault-controlled northeastern arm of Saline Valley. Travertine (calcium carbonate) and silica (silicon dioxide) precipitate out of the spring water and build into mounds. The discharged groundwater, which ranges in temperature from 102° to 112°F, became heated as it circulated to depths of about a half mile. Palm Springs, the central, hottest, and most developed of the three, discharges some 2.6 gallons of hot water per minute. The coolest spring, Upper Warm Spring, is the least developed and lies some 3 miles up the gravel road from Palm Spring.

A warm pool at the Upper Warm Spring area.

31. FATHER CROWLEY VISTA POINT
Western North American Geology Encapsulated

Without question, the sweeping view from Padre Point at the east end of Father Crowley overlook offers the most dramatic and instructive perspective of the western side of Death Valley National Park. Only 6 miles away, Panamint Valley lies some 2,500 feet below, with the bulk of the Panamint Range behind, stretching from the southern Cottonwood Mountains to beyond Telescope Peak. Immediately north, Rainbow Canyon, named for its colorful bedrock, cuts a deep gash through the basaltic lava flows that drape the east side of the Darwin Plateau. To the south, you can see the west side of the Argus Range, which reaches an elevation of 8,843 feet at Maturango Peak.

The basalt that rims Rainbow Canyon outcrops by the parking lot at the end of the gravel road. A close look shows it's very finely crystalline but contains some noticeably larger crystals of green olivine and white plagioclase. It erupted between 7.7 and 5.7 million years ago during the Miocene Epoch. In the canyon, you can see that it consists of alternating black basalt lava flows like the one at the parking lot and reddish pyroclastic, or broken explosive, material. In basaltic fields such as this one, pyroclastics typically consist of accumulations of broken rock, cinders, and ash ejected forcibly out of local vents. The red deposit along the highway about a half mile to the east and some 500 feet below is a cinder cone that is revealed by a roadcut. The red colors in both come from oxidized iron.

Looking into Rainbow Canyon from Father Crowley Vista Point, you can also see folded Paleozoic limestone intruded by Jurassic quartz monzonite, a type of granite. Together, these rocks encapsulate important events in western North America's geologic history. The limestone, part of the Permian Owens Valley Formation, lies near the top of some 30,000 feet of sediment deposited mostly in a stable ocean shelf environment that lasted from the Late Proterozoic through the Permian Period. Its folding reflects the crustal compression that affected the west coast and Death Valley afterward, from the Permian through the Cretaceous, while the granitic intrusion is a manifestation of the subduction zone and magmatic arc that replaced the stable ocean shelf.

Today's ongoing crustal extension produces volcanic activity throughout the region, evident in the basaltic lava flows, as well as the prominent dike that cuts the folded limestone. Extension also created the normal faults visible in the canyon

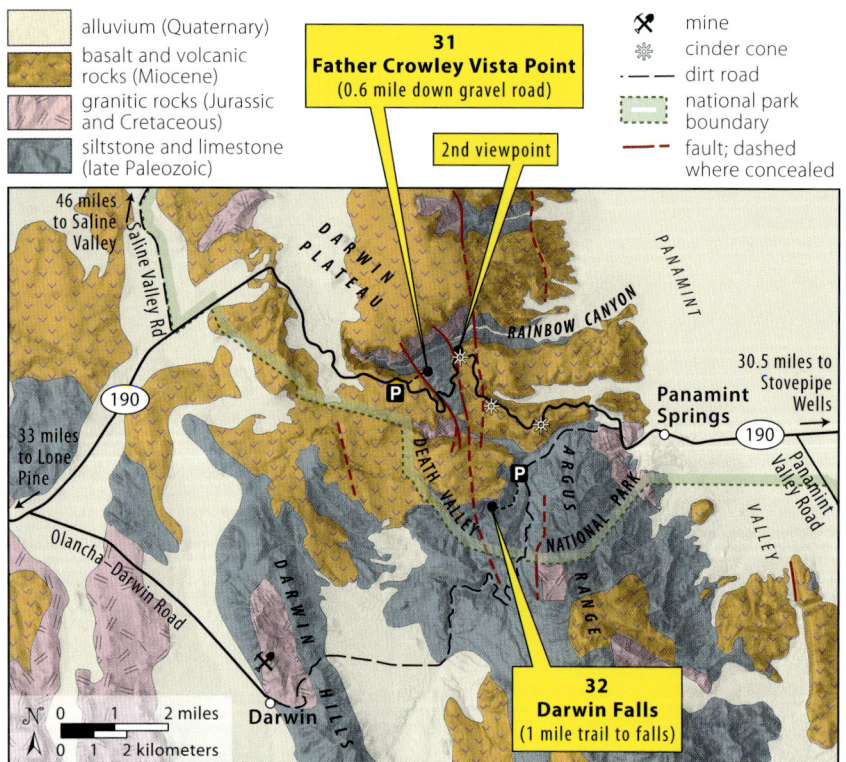

See map on page 89 for a larger overview. For the best views, drive 0.6 mile eastward down the gravel road from the paved parking lot. A second, even more spectacular viewpoint can be reached by walking to the edge of Rainbow Canyon from pullouts off CA 190 below the viewpoint.

The basaltic lava flows of Rainbow Canyon erupted between 7.7 and 5.7 million years ago.

Geologic relations in Rainbow Canyon as seen from the viewpoint accessed from a pullout along CA 190 east of Father Crowley Vista Point. Folded Permian limestone is intruded by Jurassic granitic rock and overlain by Miocene basalt, all of which is broken by an extension-related normal fault (red line). Black lines show the bedding in the limestone.

walls. Several normal faults offset the contact between the granitic rock and overlying basalt in the western part of the canyon; an especially large one offsets the basalt and limestone contact toward the east. One feature that looks like a normal fault is not a fault at all; the rocks only look offset because of the perspective afforded by the view.

Looking across northern Panamint Valley, you can see many of the same features but at a grander scale in the Panamint Range. There, Jurassic-age granitic rock of the Hunter Mountain batholith intrudes folded and thrust-faulted Paleozoic rock. You can see the Lemoigne thrust fault, active during the Jurassic Period, which placed older Cambrian and Ordovician rock over folded, younger Pennsylvanian and Permian rock. A blanket of the younger basalt overlies all the older features. The Panamint Range was uplifted by the recently active fault zone that appears at the abrupt transition between valley floor and range front. You can trace the fault zone all the way down the west side of the range.

Driving down the highway toward Panamint Valley, you initially pass through numerous outcrops of the limestone, some of which are cut by greenish, altered dikes of Jurassic age. The cinder cone lies across from a pullout after 2.5 miles. After 6.1 miles, you encounter alluvial fan deposits that underlie some of the basalt—an indication that uplift was occurring before eruption of the basalt. At 6.7 miles, you pass an exposure of the Jurassic quartz monzonite, and immediately beyond it is the turnoff for Darwin Falls (site 32).

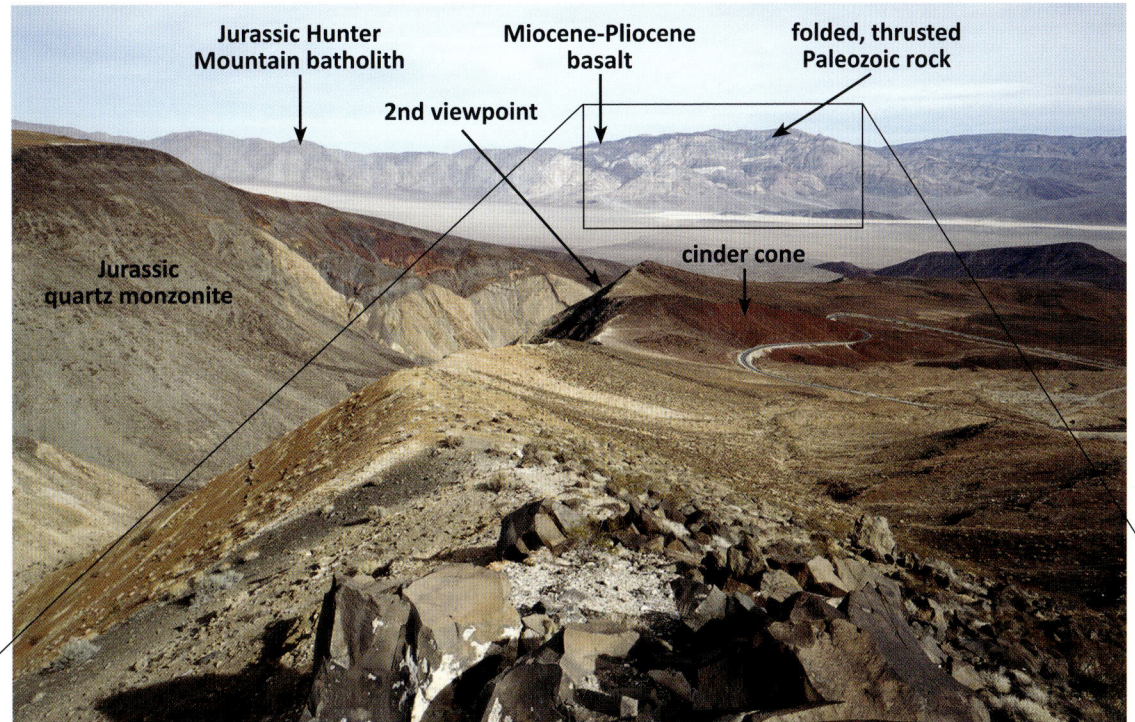

View to the northeast from the overlook at the end of the gravel road, with views of Rainbow Canyon and the northern Panamint Valley. At the base of the photo are basalt outcrops. The rectangle shows the view in the photo below.

Close-up view of Panamint Butte and its geology.

DARWIN FALLS
Not a Dry Fall
See map on page 92.

It may be hard to believe, but Darwin Falls flows year-round. Fed by a series of springs farther up the canyon, the stream continues below the falls about 1 mile before sinking into the gravel. What starts as a hike up a typical dry desert wash beginning at the trailhead off CA 190, ends after 1 mile in a green riparian wonderland of willows and cottonwoods. Studies of the water at Darwin Falls indicate it mostly originates as precipitation that infiltrates the ground at elevations above 7,500 feet and takes about 3,000 years to follow its underground route to the canyon. Because of the long travel time, the discharge remains remarkably steady. From May 2005 to September 2006, it stayed between 71 and 110 gallons per minute, excluding floods from heavy rainstorms.

Seemingly in concert with the vegetation, the rock along the hike to Darwin Falls turns green, too! The trail begins in north-tilted, brownish limestone and sandstone of the Permian Darwin Canyon Formation and ends in an endless variety of rock in varying shades of green and blue. In some places, the rock looks sedimentary while in others it looks igneous. Indeed, researchers have expressed different opinions on these rocks, but all agree the colors come from alteration by hot fluids, probably because of heating from the nearby intrusion of granitic magma during the Jurassic Period. You can get a distant view of part of the intrusion from Father Crowley Vista Point (site 31), just 3 miles to the north. The same magmatism likely resulted in much of

Darwin Falls flows over hydrothermally altered bedrock. Traces of what appear to be remnant vertical bedding in a sedimentary rock appear to the right of the upper part of the falls.

the lead, silver, and zinc mineralization that fueled the Darwin mining district, which is still an active mine less than 4 miles to the west.

Along the hike, you can also see numerous dark-colored dikes that cut through the rock. From a distance, they appear black, as if they were basalt, but up close, you can see they're fairly light colored and coarse grained, with numerous crystals of plagioclase and hornblende. Known as diorite, it likely intruded late in the Jurassic as part of the Independence dike swarm, a series of intrusions that extend more than 350 miles from the Transverse Ranges to the Sierra Nevada.

Close-up view of the rock near Darwin Falls. The green color comes mostly from the minerals diopside and epidote that can form when silicon dioxide–bearing magmas intrude and react with calcium-bearing limestones.

A diorite dike, the dark band diagonaling up to the left, intruded the Permian-age limestone (light-colored, layered rock) near the beginning of the trail. Note the young, black basalt on the skyline.

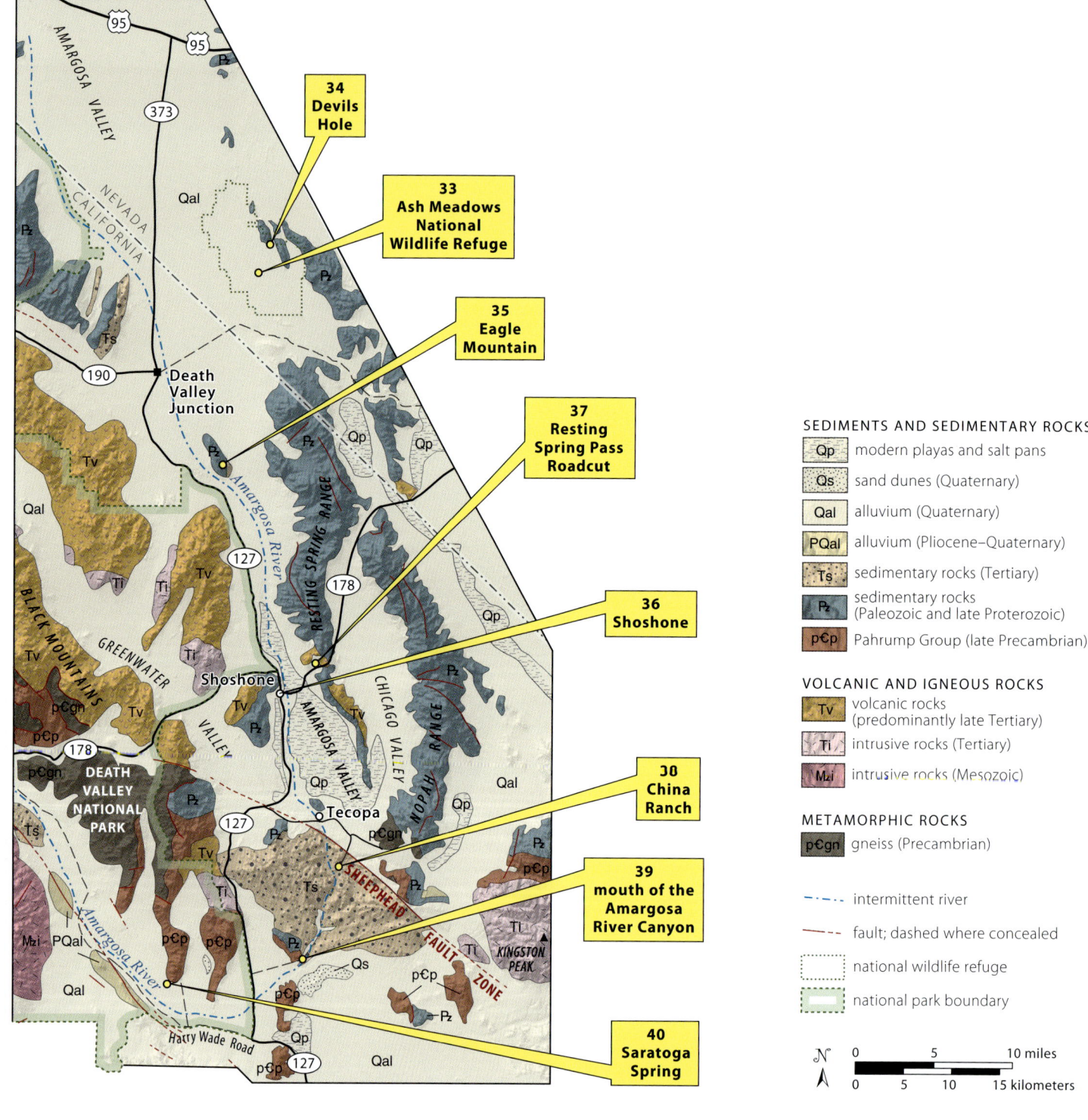

Pleistocene lakebeds fill the Amargosa Valley near Shoshone. In the background are Cambrian sedimentary rocks of the Resting Spring Range.

AMARGOSA VALLEY AND POINTS SOUTH

The Amargosa River, usually dry for most of its length except during storm events, flows southward down the Amargosa Valley, east of Death Valley and mostly outside the national park. The river water surfaces in places, creating oases in the otherwise desolate, dry region that many park visitors pass through en route from Las Vegas. Some 25 miles south of Shoshone, the Amargosa River makes a broad U-turn and empties northward into Death Valley. This connection, made in the past 300,000 years, links the two valleys physically and hydrologically. Several of the sites, located where the water comes to the surface, speak to the hydrologic connection and to a lake that occupied much of the valley in Pleistocene time.

    The bedrock that protrudes above the valley floor pertains to Death Valley National Park's more ancient history. Exposures of the Kingston Peak Formation at the mouth of the Amargosa Canyon (site 39) point to ice age conditions that prevailed over much of the planet during the Proterozoic Era, between about 720 and 635 million years ago. Tilted fault blocks and volcanic activity display the crustal extension that began about 14 million years ago. China Ranch Basin formed about 12 million years ago along the Sheephead fault zone, a major strike-slip fault that parallels the Furnace Creek fault zone to its north.

ASH MEADOWS NATIONAL WILDLIFE REFUGE
Oasis in the Mojave Desert

The abundant spring water in Ash Meadows National Wildlife Refuge, the largest remaining oasis of the Mojave Desert, is sourced about 100 miles away in Nevada's Pahranagat Mountains. It takes at least a few thousand years (estimates range up to 30,000 years) for the water to make the trip, but its presence supports a wide variety of plant and animal life, including four species of pupfish that have lived here since the ice ages.

Most of the more than thirty springs and seeps rise along the Ash Meadows fault zone, which separates the low-lying alluvial and lake deposits of the Ash Meadows area from uplifted limestone and dolomite bedrock to the east. The springs' combined discharge is approximately 10,000 gallons per minute. On its own, Crystal Pool, just 2 miles west of Devils Hole, discharges more than 2,800 gallons per minute. In the 1970s, several human-made reservoirs, including Crystal Reservoir, were built to hold the water for irrigation purposes, but, with the creation of the national wildlife refuge in 1984, are no longer needed.

Low sand dunes, stabilized by vegetation, occupy much of the western edge of the refuge. They act as natural dams, causing the spring-fed surface water to back up and support marshes and wetlands. Ash Meadows National Wildlife Refuge maintains a visitor center and three accessible boardwalks to some of its springs, including Crystal Pool. The boardwalks range in length from 0.11 to 0.75 mile. It also hosts the Ash Meadows Fish Conservation Facility, which breeds and houses Devils Hole pupfish.

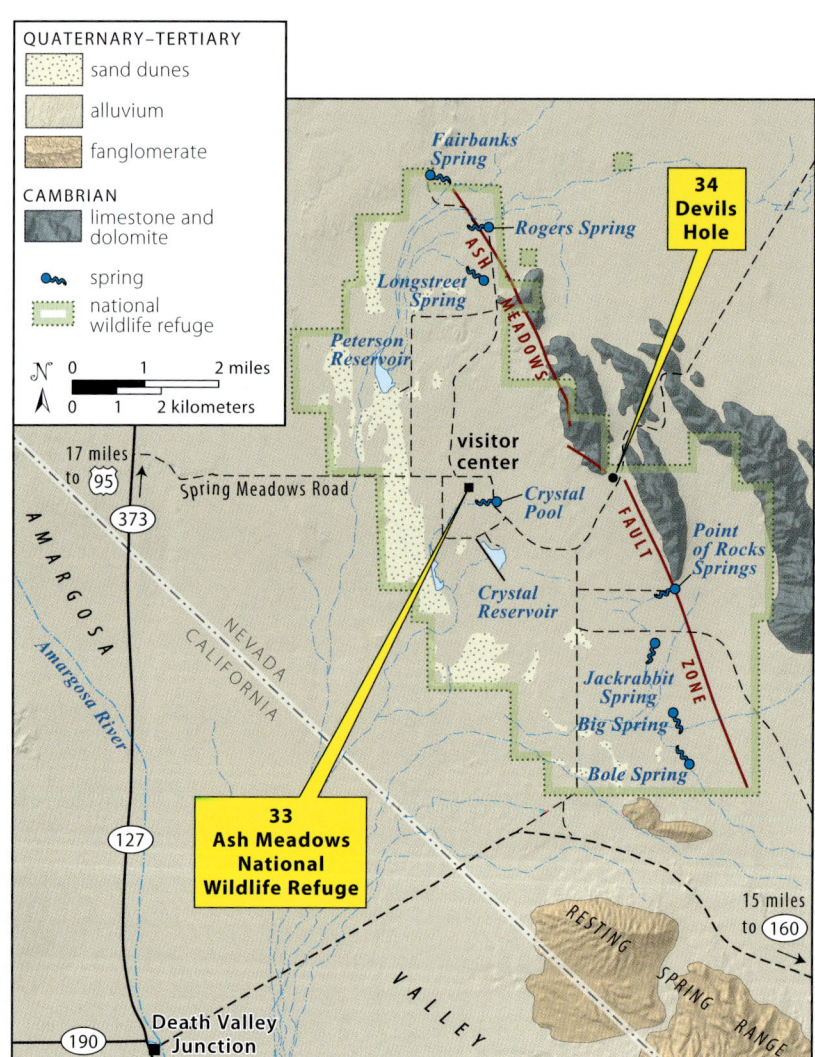

Springs at Ash Meadows National Wildlife Refuge mainly lie along the Ash Meadows fault zone.

Kings Pool at the Point of Rocks spring complex. The ridge in the background lies on the uplifted side of the Ash Meadows fault zone.

Mud cracks form along the drying shore of Crystal Reservoir, which was constructed in 1971 for irrigation.

The opening to Devils Hole likely formed about 60,000 years ago when the roof to the fissure system collapsed. The National Park Service constructed the enclosure and monitors the area to protect the habitat from vandalism. The hills in the background rise along the Ash Meadows fault zone.

DEVILS HOLE
Pupfish Survive in Diabolic Pool
See map on page 100.

Protected as a separate 40-acre parcel of Death Valley National Park within Ash Meadows National Wildlife Refuge, Devils Hole angles into the limestone bedrock as a fissure along a fault zone. About 50 feet down, it reaches a narrow pool of water measuring some 11 by 72 feet—one of those unusual spots where groundwater daylights on its subterranean journey. Nobody knows how deep the fissure actually goes, but divers mapped a similarly narrow cave system that extends at least 500 feet below.

The 1-inch-long Devils Hole pupfish (*Cyprinodon diabolis*) lives in the upper 80 feet of water at Devils Hole. This pupfish lives nowhere else, making it the most geographically restricted vertebrate species on Earth. As if things couldn't be more difficult for the fish, the water maintains the unusually warm temperature of 93°F and is poorly oxygenated. Moreover, the fish's main feeding and spawning area consists of a flat limestone shelf just below the water surface at the south end of the pool. With this tiny habitat, a lifespan of only 10 to 14 months, and a population of only a few hundred members at most, the fish is highly vulnerable to environmental changes, especially decreases in the water level.

In 1967, the fish earned a spot on the original Endangered Species Preservation Act, which was tested when nearby groundwater pumping started lowering the water level and threatened

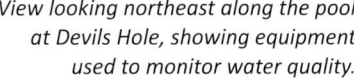
Banded calcite vein next to the pool. Pen for scale.

View looking northeast along the pool at Devils Hole, showing equipment used to monitor water quality.

to expose the shelf. In 1976, the case went all the way to the Supreme Court, which ruled in favor of the fish. Since then, the fish population rebounded, but it continues to fluctuate. In 2006 and 2013, it fell to lows of 38 and 35 individuals, respectively; in the fall of 2022, it reached a high of 263. The National Park Service constructed fencing and a viewing platform to protect the pool and, along with the National Fish and Wildlife Service, continually monitors the water level in the pool and conducts twice yearly population counts.

It's not clear to researchers when the pupfish first colonized the spring, but they are sufficiently different from nearby pupfish, suggesting they've been isolated for at least tens of thousands of years. There's also an important clue beneath the water surface. Deposits of calcite, more than 1 foot thick in many places, line the fissure walls; radiometric dating of the layers indicates they were precipitated continuously from about 560,000 to 60,000 years ago. Most researchers agree calcite precipitation ceased because of changing water conditions, likely caused by collapse of the cave ceiling, which would have opened the spring to the outside world and new inhabitants. This scenario poses a conundrum,

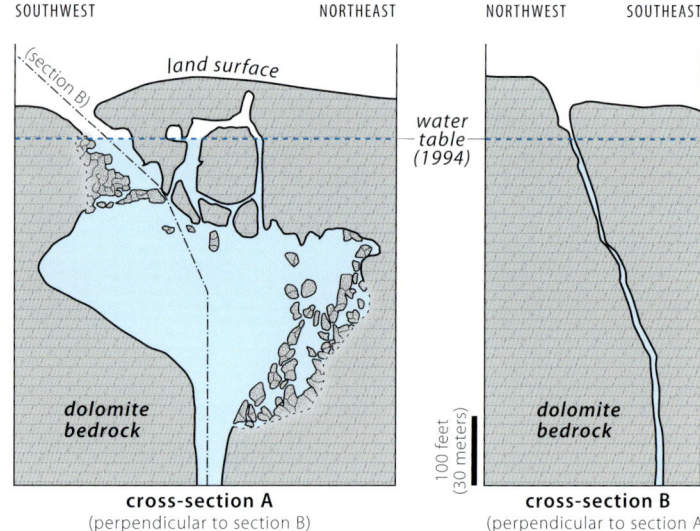

Mutually perpendicular cross sections of the cave system of Devils Hole. As shown by cross section B, the northwest-southeast dimension of the cave is very narrow, but it widens considerably in the southwest-northeast orientation as shown by cross section A.
—Cross sections modified from Riggs and Deacon, 2002

Pupfish on the algae-covered shelf at Devils Hole. —Photo by Olin Feuerbacher, US Fish and Wildlife Service

however, because geologists find no evidence for a surface water connection from Devils Hole to any other surface water sources within that time frame.

By measuring oxygen isotopes in the calcite layers, researchers have been able to document temperature changes over the past 500,000 years. Most long-term climate studies are based in studies of ice cores or marine sediments, so Devils Hole provides a valuable land-based test of this record. Similar to the other studies, the researchers found the area experienced four major periods of cooling during this time period that correspond to glacial advances at higher elevations and latitudes.

Seemingly without warning, the water at the surface of Devils Hole can slosh wildly around the pool in response to certain distant earthquakes. These events, called seiches, can last as long as a half hour and cause the water to splash 4 to 5 feet above its resting level. The seiches likely result when low-frequency seismic waves, undetectable by humans, travel at the right frequencies, strengths, and trajectories to squeeze and extend the narrow fissure below the small pool, causing the water to flow in and out of the pool. Seiches have been observed at Devils Hole on at least three occasions, most recently in April 2012 in response to a magnitude 7.4 earthquake in Oaxaca, Mexico.

View looking northeast from the viewing bridge. The pool lies along a small fault zone (dashed red line) in the Bonanza King Formation.

35 EAGLE MOUNTAIN
Soaring Landmark of Tilted Cambrian Beds

Eagle Mountain rises straight out of the middle of the Amargosa Valley, and you get full-on views of both its south and west sides from CA 127. The west side beautifully displays the Cambrian sedimentary sequence from the Zabriskie Quartzite up through most of the Bonanza King Formation. The Zabriskie Quartzite, the pink beds at the bottom of the mountain, consists of a nearly pure quartz sandstone that was deposited in coastal environments. Above it, the Carrara Formation consists of brown, gray, and tan beds of siltstone, shale, and limestone deposited in shallow ocean waters near the continental edge. Erosion of the continent provided the silty material. Dolomite and limestone of the Bonanza King Formation, deposited farther offshore, forms the gray cliffs that reach to the top of the mountain. Together, these three rock units mark the encroachment of a shallow ocean onto the land, brought about by subsidence of the continental shelf throughout much of the Cambrian Period.

From the west side, you can see several normal faults, especially toward the north end of the mountain where the Carrara Formation drops to progressively lower elevations. You can also pick out a series of light-tan beds within the middle part of the Bonanza King

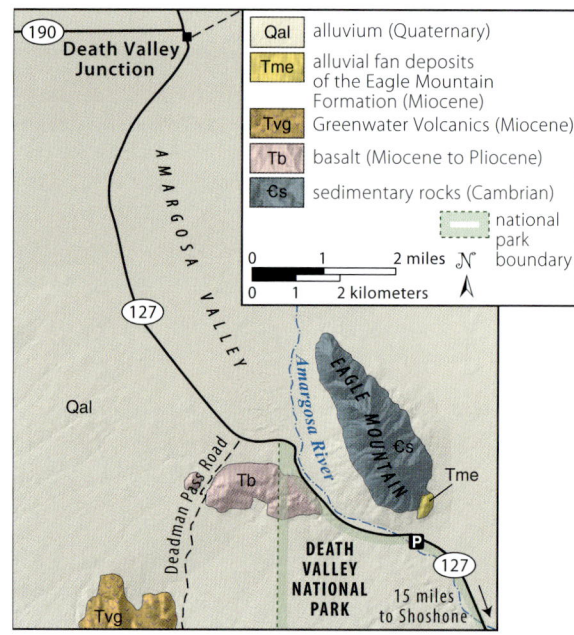

The easiest access point to view the Eagle Mountain Formation on the mountain's base is at the pullout where CA 127 crosses the Amargosa River.

View of Eagle Mountain from CA 127 looking east. The pink outcrops low on the slope and on the right side of the photo are Zabriskie Quartzite. The brown-and-tan-striped beds above the Zabriskie are the Carrara Formation, and the dark-gray and tan rocks above them in the upper half of the mountain are the Bonanza King Formation.

Aerial view looking southward over Eagle Mountain and down the Amargosa Valley.

Formation that shows the faulting. In one place, this tan marker is cut by three parallel normal faults, mimicking the style of tilted fault blocks across the Basin and Range. Notice how each block is rotated down and into the fault that extends beneath it.

From its south side, you can see that Eagle Mountain is a tilted fault block, defined by steeply east-tilted beds that rotated toward the fault that runs up the east side of the Amargosa Valley. The Resting Spring Range, which is uplifted along that fault, consists of Precambrian to Cambrian rocks that also tilt eastward. Similarly, another normal fault must exist on the west side of Eagle Mountain to separate it from the younger volcanic rock of the Greenwater Range to the west.

The low-lying, yellowish rocks on the southeast edge of the mountain, called the Eagle Mountain Formation, were deposited by rivers between about 15 and 11 million years ago. They're also tilted but not quite as much as the Bonanza King Formation, indicating some tilting occurred prior to their deposition. The Eagle Mountain Formation contains rounded, gravelly pieces of granitic rock that came from the Hunter Mountain batholith more than 60 miles to the northwest.

Site continues ⟶

Eagle Mountain from the south, showing the east-tilted nature of the bedrock. The yellowish rocks on the right side of the photo are the Eagle Mountain Formation, which extends partway up the right side of Eagle Mountain.

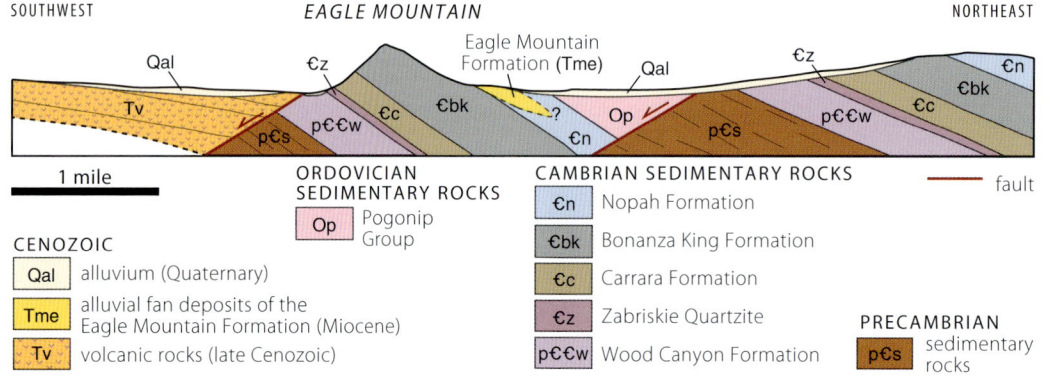

Schematic cross section of Eagle Mountain showing the faults on its southwest and northeast sides. The exact placement of the faults is speculative.

Telephoto view of normal faults at the northwest end of Eagle Mountain. Yellow arrows point to individual faults, which slope down to the left.

36 SHOSHONE
Pleistocene Lakebeds Entomb Ash and Fossils

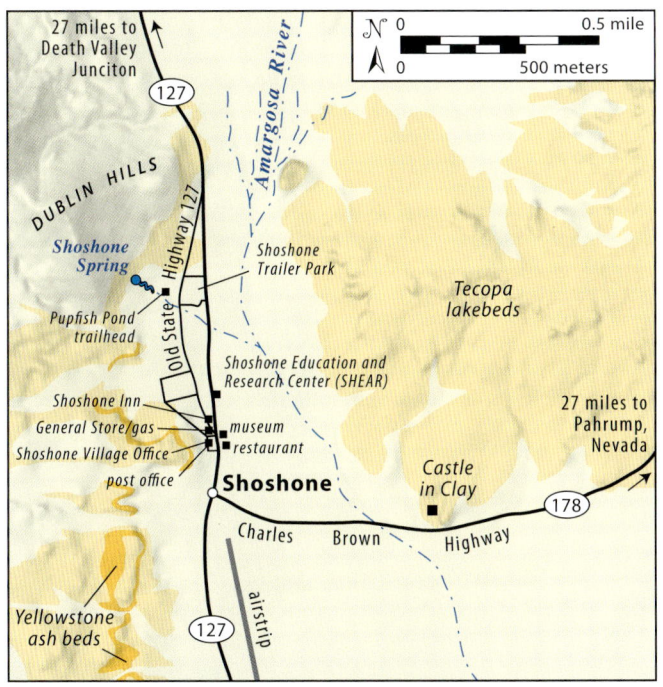

Location map for Shoshone, California, on CA 127, just north of the intersection with CA 178.

No story of Death Valley geology would be complete without including features in the Amargosa Valley near the small town of Shoshone. Here, astride the Amargosa River, deposits of Lake Tecopa attest to wetter conditions during the Pleistocene ice ages. The Tecopa lakebeds fill the low areas from just north of Shoshone to the south end of the town of Tecopa, an area of about 50 square miles. The lakebeds also include numerous alluvial fan, river, and windblown deposits, as well as spring deposits. The white-to-tan layers of mudstone, sandstone, conglomerate, and even volcanic ash preserve an outstanding record of the area's landscape and climate from just before 2.4 million years ago to 300,000 years ago. Erosion has carved the lakebeds into a maze of buttes, small mesas, and rounded hills that are especially striking just east and south of town.

Numerous researchers, most notably Dan Larson of the University of Memphis and Marith Reheis of the US Geological Survey and their colleagues, have put together detailed studies of the lake basin with the benefit of ages obtained from volcanic ash layers that settled to the lake bottom. Until about 1 million years ago, the basin likely consisted mostly of small brackish lakes and mudflats with alluvial fans along the edges. The first deposits of the Amargosa River appear in the record of the northern part of the basin

View looking east over Shoshone. The Tecopa lakebeds form the low cream-colored hills in the middle ground while Cambrian sedimentary rocks of the Resting Spring Range form the background.

between about 1 million and 765,000 years ago, when the Bishop Ash was erupted from the Long Valley caldera in Owens Valley. Water carried by the river expanded the lakes, which merged into a single large lake at various wetter times and shrunk into isolated smaller lakes during arid periods. The lake drained southward and eventually into Death Valley after Amargosa Canyon—just south of Tecopa—was eroded, probably between about 300,000 and 140,000 years ago.

The lakebeds are well known for their fossilized mammals, including mammoths, five species of camels, llamas, horses, and a variety of rodents. Some localities even contain fossil flamingos. Many of the camels were discovered near Tecopa with their limbs in the standing position, suggesting the animals died after becoming mired in mud-rich springs. Other animals, found closer to Shoshone, are preserved in giant vertical fractures that cut the sediments, which suggests they died after falling into them.

One fossil mammoth, at least 650,000 years old, fills a 20-foot-long display case in the Shoshone Museum. The many other displays include relics of the region's human and mining history as well as its geology. The museum also hosts an outdoor interpretive display designed and created by Bennie Troxel, one of Death Valley's long-time geologists. The large rock samples consist of Precambrian basement gneiss through the Pahrump Group and overlying Paleozoic sequence and all the way to the volcanic rocks exposed at Resting Spring Pass (site 37). A sign next to the display shows the region's stratigraphic section and interprets the Cambrian sedimentary sequence beautifully exposed on the Resting Spring Range east of town.

If you contact Shoshone Village first, you can get directions and permission to walk to a small, abandoned pumice quarry on private land in the lakebeds on the west side of town. The quarry walls display irregularly shaped, folded volcanic ash deposits from

Mammoth exhibit in the Shoshone Museum.

the Yellowstone eruption of 630,000 years ago, more than 600 miles away. The ash bed, more than 3 feet thick, contains a substantial amount of sand and silt and is locally cross-bedded, indicating redeposition by streams. While the ash demonstrates the enormous scale of the eruption, the folds are likely products of local earthquake events that shook the area when the deposit was still unconsolidated and saturated with water. In many places, you can see vertical zones where fluids escaped rapidly upward and others where denser parts of the ash bed abruptly sank downward into less-dense material. Notice how neither the underlying nor overlying beds are folded, indicating the folded beds must have been soft when they deformed and the deformation occurred before deposition of the overlying material.

Near the north end of town, Shoshone Spring hosts thousands of pupfish, all of which are a single species that is distinct from those found in Tecopa, Death Valley, and Devils Hole. Between about 1970 and 1986, this species (*Cyprinodon nevadensis shoshone*) was considered extinct because no members were found in surveys conducted during that time. However, with the discovery of numerous pupfish downstream from the spring in 1986, the town of Shoshone rehabilitated the spring area. Today, the spring supports a thriving population of the fish in a series of pools that can be viewed from several well-maintained trails winding through otherwise thick riparian vegetation. The area also hosts habitat for the Amargosa vole, an endangered species that survives only in the region. A kiosk and interpretive signs mark the entrance to the area.

Once considered extinct, the Shoshone pupfish now thrive near Shoshone Spring. The fish are about 1 inch long.

Deformed beds of Yellowstone ash in the quarry on the west side of town, accessible with permission from Shoshone Village.

ROADCUT AT RESTING SPRING PASS
Gas Bubbles Tilted by Faulting

Better than any other single rock exposure, this roadcut illustrates the story of crustal extension in the Death Valley region. Park in the large pullout on the south side of CA 190 near the crest of the Resting Spring Range, 4 miles east of Shoshone. The most-striking feature of the roadcut, the prominent black stripe on the west side, is highly fractured obsidian, highly compacted volcanic glass. It lies near the bottom of an ash flow, a mixture of hot gases and ash that probably erupted from a now-buried caldera about 10 million years ago. Notice how the ash-flow tuff on either side of the obsidian becomes increasingly welded, denser, harder, and darker as you approach the obsidian. Such obsidian zones can form near the hot bases of ash-flow tuffs because the overlying material causes them to compress and weld together. The rock becomes increasingly dark as tiny gas bubbles, which reflect light, get squeezed out.

A closer look at the obsidian shows much of it contains large, elliptical-shaped gas bubbles—some more than an inch long—formed by the coalescence of many tiny bubbles. The bubbles

In the roadcut at Resting Spring Pass, the black stripe is a zone of obsidian (volcanic glass) formed near the base of the 10-million-year-old ash-flow tuff. Notice the faults and the large gray limestone block of the Bonanza King Formation within the lakebeds seen on the right side of the roadcut.

are flattened because of the compaction and are all oriented the same, showing a gentle eastward inclination. Considering flattening of the bubbles would occur directly downward because of the compaction, their current orientation must be the result of eastward tilting, one of the hallmarks of the area's modern crustal extension. The obsidian zone itself slants westward—in the opposite direction as the gas bubbles—because of the ancient topography. When the ash flow erupted, it moved partway up a west-facing slope before coming to rest on older rocks.

Just beneath the ash flow, the rocks consist mostly of older ash-flow tuff in varying shades of pink and orange, along with some tuff-rich river gravels and lakebeds. These rocks are tilted the same amount as the gas bubbles in the obsidian. Two west-dipping normal faults that formed during the extension offset these rocks in a sense that's compatible with the tilting. Perhaps

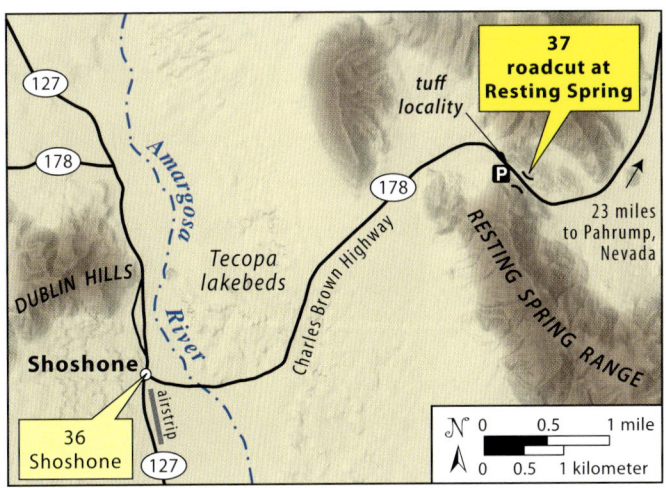

Resting Spring Pass, at an elevation of about 2,340 feet, marks the divide between the Amargosa Valley to the west and the Chicago Valley to the east.

Close-up view of one of the normal faults exposed in the roadcut. The rock hammer rests on ash-flow tuff that predates the 10-million-year-old tuff. The gravel on the right side, underneath a yellow tuff, was deposited on an alluvial fan, the presence of which indicates substantial topographic relief existed before deposition of the overlying rocks. Thicker deposits of the gravel outcrop on the south side of the parking area.

Obsidian zone near the base of the ash-flow tuff. Note how the surrounding ash-flow tuff grows darker in color toward the obsidian from both sides, brought on by increasing degrees of welding and compaction of the rock.

Flattened air bubbles in the tuff are tilted back to the east.

the most-significant feature of the older rocks, however, is the large block of Cambrian limestone that looms over the lakebeds at the top of the eastern side of the roadcut. Part of a large landslide block, it slid into a lake that formed over the top of the older ash-flow tuff. The block indicates the nearby presence of substantial topographic relief before the tilting occurred. Still older gravel, consisting mostly of particles from the Bonanza King Formation, outcrops beneath road level on the south side of the parking area.

Looking southward along the front of the Resting Spring Range, you can see evidence for an older episode of faulting that likely caused this relief. There, the Cambrian rocks are tilted even more steeply than the volcanic rocks of the roadcut, indicating a period of extension and tilting that took place before the younger rocks were deposited.

38 CHINA RANCH
A Date with Another World

When you drive to China Ranch, a working date farm south of Tecopa, it feels like you're entering a parallel universe the moment the road begins its steep descent to China Ranch Creek. The access road traverses dry desert then drops down through walls of fanglomerate until the valley opens up and you are greeted with lush, green vegetation. Rows of date palms are irrigated with the plentiful water from Willow Spring, about 1 mile up the canyon. China Ranch offers informative displays about the natural and human history of the area, hiking trails, a picnic area, and a gift shop with a variety of date products, including shakes.

Surrounding the green oasis are spectacular badlands eroded into light-colored sedimentary rocks deposited between about 12 and 8 million years ago in the China Ranch Basin. In addition to the fanglomerate along the road, the basin rocks consist mostly of mudstone—parts of which contain abundant gypsum—as well as thinly bedded siltstone and sandstone. The rocks also include megabreccia—deposits containing large, angular blocks, likely from rock avalanches—and even some small patches of freshwater limestone. The badlands were eroded into the finer-grained, impermeable rocks because the lack of water infiltration

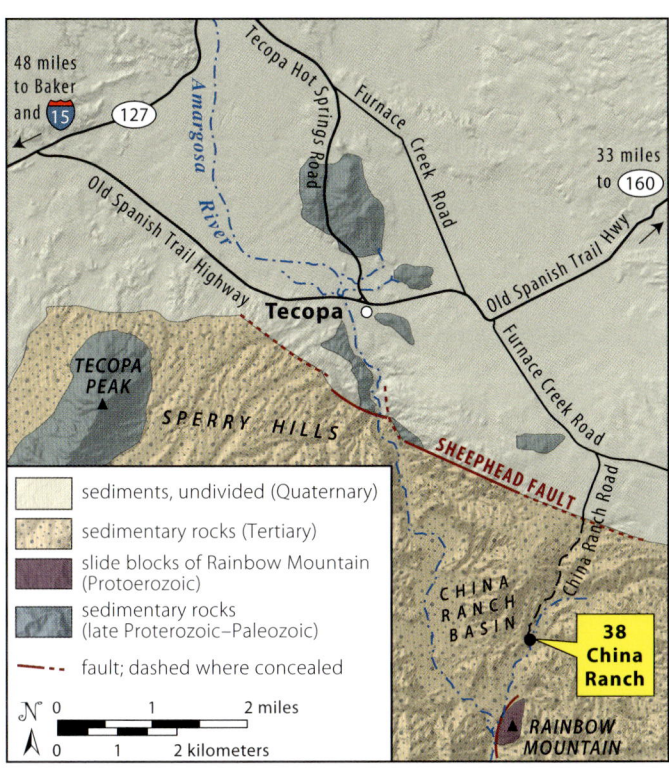

China Ranch is just over 5 miles south of Tecopa by road, the last 2 miles on a gravel road.

View to the northeast up China Ranch Creek over the trailhead, parking area, and date farm of China Ranch. The lighter-colored beds are mostly mudstones and siltstones, while the darker hills consist of fanglomerate.

discourages vegetation. Clays, which expand when wet, create a smooth coating over the sedimentary rock.

The extent of the various sedimentary rocks outlines the somewhat elongate shape of the China Ranch Basin. It trends approximately east-west, with its northern limit defined by the Sheephead fault zone. In general, the fanglomerate—deposited in alluvial fans as channel deposits—formed the basin's edges, along with megabreccia units that cascaded down from elevated areas. The finer-grained rocks—deposited in sandflats, salt-rich mudflats, and both ephemeral and semi-permanent lakes—occupied the middle. The abundance of gypsum near the center of the basin suggests it was a closed basin; water could flow into the basin but not out, leaving only through evaporation. Springs issuing from the toes of the alluvial fans likely deposited the limestone.

If you hike 1 mile downstream from the ranch to the confluence of China Ranch Creek and the Amargosa River, you'll encounter Rainbow Mountain, a colorful jumble of blocks that slid into the basin during its early history. Individual blocks consist of the brecciated Crystal Spring Formation, Beck Spring Dolomite, Noonday Dolomite, and Johnnie Formation—all Proterozoic in age. The blocks of old rock are interleaved with the much younger rocks of the China Ranch Basin.

Fanglomerate of the China Ranch Basin along the gravel access road.

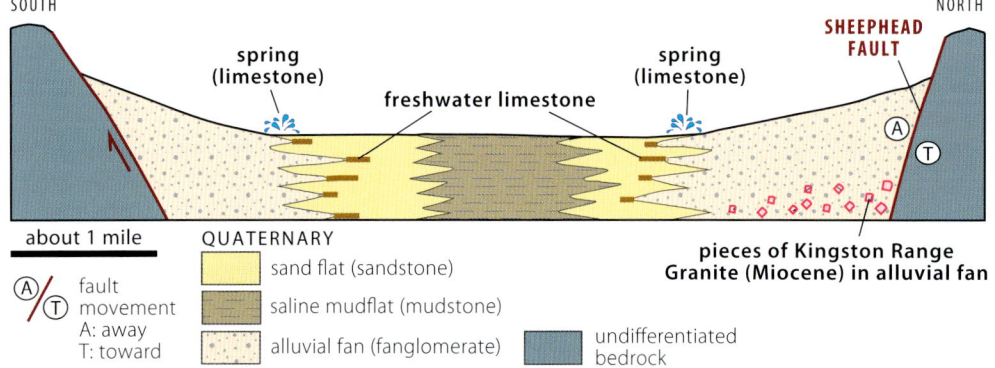

Cross section through the China Ranch Basin showing bounding faults and material shed into the basin off the rising mountains. The Sheephead fault on the north side likely moved right-laterally, moving the basin away from its original source area south of the Kingston Range, which is currently north and east of the basin. The pink shapes at the base of the fanglomerate on the right side of the cross section represent granitic sediments derived from the Kingston Range Granite.
—Cross section modified from Scott, 1985

Christopher Davidson of San Francisco State University found that movement on the Sheephead fault on the northern edge of the basin coincided with the formation of the basin and controlled much of its subsequent evolution. Most notably, he found the conglomerate near the bottom of the sequence contained pieces of granite derived from the Kingston Range, about 15 miles to the southeast. These granite cobbles are absent from younger parts of the conglomerate, suggesting that as the fault moved, it displaced the basin away from its granite source.

Rainbow Mountain, at the confluence of China Ranch Creek with the Amargosa River, consists of blocks of Proterozoic rock that slid into the basin from nearby mountains. Two patches of light-colored beds, one halfway up the slope and the other below the right skyline, are lakebed deposits that are interbedded with different landslide deposits. On the left skyline are lakebed deposits that were deposited over the top of all of them, but if you walk into the canyon on the left (arrow), you can see lakebed deposits beneath all of them!

Lakebeds of the China Ranch Basin.

The tan-colored Noonday Dolomite overlies the reddish Kingston Peak Formation near the mouth of the Amargosa River Canyon.

MOUTH OF THE AMARGOSA RIVER CANYON
Iceberg Dropstones of Snowball Earth

Imagine so much of our planet covered by glaciers that there was sea ice near the equator. That was the picture some 716 to 660 million years ago when the red rocks at the mouth of the Amargosa River Canyon were deposited. Glaciation was so widespread that geologists call the episode Snowball Earth. These rocks, along with similar rocks of the same age found on all of today's continents, contain evidence of the ice.

The red rocks in the upper Kingston Peak Formation consist mostly of shale, with siltstone, sandstone, minor amounts of conglomerate, and even some limestone—all deposited in an ocean. Many of the beds are turbidites, which grade from coarse-grained sand at their bases to shale at their tops, suggesting deposition in deep water from turbid, sediment-charged currents flowing down the continental shelf. The Kingston Peak was deposited along the edge the supercontinent Rodinia as it was in the process of breaking up. Normal faults were active during the rifting event and contributed to variations in the formation's thickness. At this site, for example, it measures about 720 feet in thickness, but it's more than 10,000 feet thick in the nearby Kingston Range.

The beds that most directly suggest glaciation consist of thinly layered shale or siltstone that encloses isolated cobbles or boulders. These rocks posed an enigma to early geologists

118

because shale and siltstone are deposited in quiet water, the energy of which is too low to transport the boulders. Geologists eventually determined the rocks were carried by icebergs that drifted over quieter parts of the seafloor. As the icebergs melted, they released their payload of boulders, known as dropstones, to settle onto the fine-grained ocean sediment. Some dropstones display glacial striations, scratches caused by ice-entrained rocks as glaciers scour bedrock. Some deposits at this locality consist of numerous rock fragments enclosed in a fine-grained matrix of mud or silt, resembling glacial till. These deposits likely originated as till but were then redeposited by submarine debris flows in front of the ice.

It doesn't take a lot of iron oxide in a rock to color it red, but some of the thin red beds here are true iron stones, with iron oxide contents exceeding 75 percent by weight. Such deposits are rare in rocks younger than about 1.8 billion years old when Earth's atmosphere became sufficiently

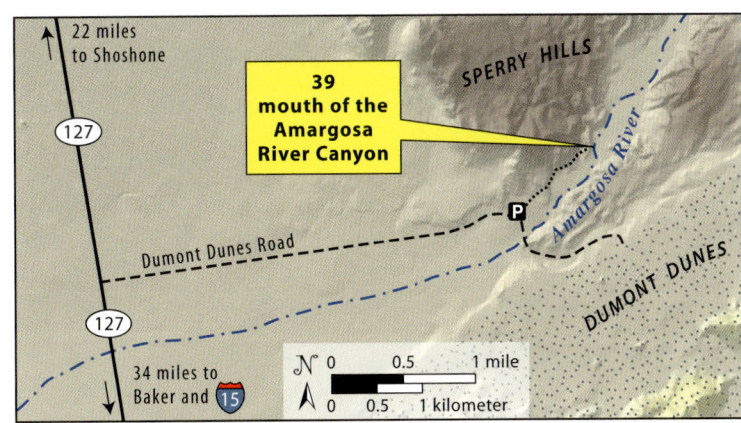

The site lies just upstream from the mouth of the Amargosa River Canyon at Dumont Dunes. Park your car in a pullout near the parking symbol and walk up the sandy 4WD road to the ridge marked by the yellow pointer. The rock there consists of red shale, siltstone, and glacial till–like deposits and contains numerous dropstones.

This outcrop of the Kingston Peak Formation is considered to be glacial till that was remobilized as a debris flow and deposited underwater.

One of the larger dropstones, a boulder of the older Beck Springs Dolomite, is surrounded by fine-grained ocean deposits. Look for it at the east base of the outcrop at its north end.

oxygenated, preventing iron from dissolving and being transported to a depositional site. Iron-rich sedimentary rocks, however, are found in association with glacial deposits of the same age in many places around the globe. Some researchers argue the ice sheets created temporary low-oxygen conditions by covering large tracts of ocean. When parts of the sheets melted and flooded these areas with oxygenated water, the dissolved iron precipitated.

The tan-colored cliffs above the Kingston Peak Formation consist of carbonate rock of the Noonday Dolomite. According to proponents of the Snowball Earth hypothesis, the carbonate precipitated in the ocean environment during a period of rapid warming that caused the glaciers to melt.

South end of the outcrop. The dropstones are easiest to find along its east (right) edge at the far end.

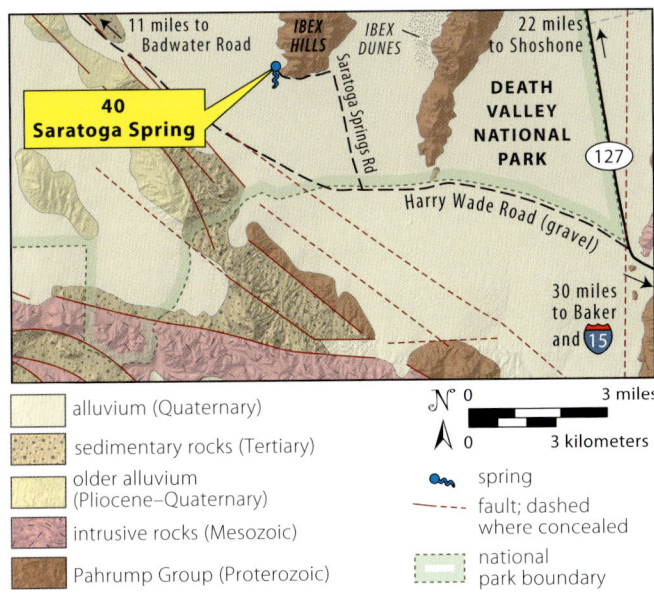

- alluvium (Quaternary)
- sedimentary rocks (Tertiary)
- older alluvium (Pliocene–Quaternary)
- intrusive rocks (Mesozoic)
- Pahrump Group (Proterozoic)
- spring
- fault; dashed where concealed
- national park boundary

Saratoga Spring is accessed via a well-graded gravel road from CA 127. From CA 127, take Harry Wade Road west for 6 miles to the Saratoga Springs Road, head north for a little more than 2.5 miles, and then turn left and drive west 1.3 miles along the base of the hills.

SARATOGA SPRING
Solitude Amidst Former Talc Mines

Accessed by a gravel road from CA 127, Saratoga Spring is an oasis in an otherwise remote location at the southern edge of Death Valley National Park. With Death Valley's extreme evaporation rates, it's a wonder that the spring can maintain two pools, the largest of which exceeds 4 acres. The steady recharge of water comes mostly from underground flow originating from the Amargosa River near Dumont Dunes. The groundwater percolates through the gravel along the course of the dry river and into the slight topographic depression that hosts the spring. Deeper groundwater that works its way to the surface along the fault zone at the base of the nearby Ibex Hills also contributes to the spring. This deep groundwater originates as rainwater and snow in the Spring Mountains of Nevada. Occasional flood events spill into the small basin, causing the ponds to temporarily expand and bolster the water supply to the adjacent wetlands.

The thriving wetlands includes its own species of pupfish (*Cyprinodon nevadensis nevadensis*), isolated from the numerous

The main pond of Saratoga Spring with the southern Ibex Hills in the background. The rock units consist of the Crystal Spring Formation (C), Horse Thief Springs Formation (H), and Beck Springs Dolomite (B). Diabase (d) intruded the top of the Crystal Spring Formation but not the much-younger Horse Thief Springs Formation. The white coloration along the haulage road is talc.

Aerial view of Saratoga Spring looking northeast.

other species by the drying climate at the end of the Pleistocene Epoch. The habitat also supports numerous bird species and a host of invertebrates, including a June beetle and two snail species endemic to the Amargosa River drainage.

The southern Ibex Hills to the east of the pools is a block of well-layered sedimentary rocks that rises abruptly behind a normal fault. From the valley floor to the ridgetop, the layers consist of dolomite of the Crystal Spring Formation, sandstone and limestone of the Horse Thief Springs Formation, and Beck Spring Dolomite, all of which belong to the Proterozoic-age Pahrump Group. The dark-greenish rock in the lower half of the sequence consists of diabase, an intrusive igneous rock with a chemical composition similar to basalt. It intruded roughly parallel to layering in the Crystal Spring Formation about 1,050 million years ago. The overlying Horse Thief Springs Formation is significantly younger, with a maximum age of about 790 million years.

A haulage road and several former talc mines dot the edges of the diabase. Some of the mines are particularly noticeable

Exposure of talc in dolomite of the Crystal Spring Formation along the haulage road that climbs into the Ibex Hills. Note the black diabase and overlying Horse Thief Springs Formation in the left background.

You can find stromatolites, fossilized cyanobacteria, in dolomite of the Crystal Spring Formation. They were some of the only life forms that existed 1 billion years ago when this unit was deposited in shallow seawater.

because of the bright-white talc at the surface. The mineral formed as the result of chemical exchanges between the hot, intruding diabase magma and the dolomite host rock. Talc, an amazingly versatile mineral critical to industrial components, was second only to borate mining when it came to profitable mining ventures in Death Valley. Former talc mines exist in Death Valley only where the diabase intruded dolomite of the Crystal Spring Formation, such as in the Black Mountains as far north as Badwater Road and in the southern Panamint Range.

GLOSSARY

alluvial fan. A low, fan-shaped, gently sloping deposit of alluvium bordering the base of a steep slope at the mouth of a canyon.

alluvium. The unsorted, unconsolidated boulders, cobbles, gravel, sand, and finer rock debris deposited in relatively recent geologic time, principally by running water. Adjective: alluvial.

andesite. A dark-colored, fine-grained volcanic rock intermediate in composition between a rhyolite and basalt.

anticline. A fold in layered rocks with stratigraphically older rocks in its core. Most anticlines are convex upward in shape.

ash. *See* volcanic ash.

ash flow. A hot mixture of volcanic gases and ash that travels as a density current down the flanks of a volcano or along the surface of the ground.

ash-flow tuff. A tuff deposited from an ash flow.

badlands. Barren, steep, intricately dissected topography developed by erosion in fine-grained but coherent sediments or sedimentary rocks.

bajada. A broad, continuous alluvial slope extending from the base of mountain ranges out into and around an inland basin, formed by the lateral coalescence of a series of separate alluvial fans.

basalt. A fine-grained, dark, extrusive igneous rock, relatively rich in iron, calcium, and magnesium, and relatively poor in silicon.

basement. The deepest exposed crustal rock, or complex of crustal rocks in a given area.

basin. A broad enclosed depression, sometimes without drainage to the outside.

Basin and Range. A physiographic province in the western United States characterized by long, linear, fault-block mountains separated by intervening valleys.

batholith. A mass of coarse-grained granitic igneous rock, initially intruded at depths of 2 to 12 miles in the continental crust, and now exposed over a broad area and consisting of two or more plutons.

bedding. The principal layering of sedimentary rocks, typically defined by changes in grain size or color.

bedrock. Relatively solid rock that remains in its original position. It may lie beneath soil or loose rock detritus.

boulder. A rock fragment larger than 10 inches diameter. It may be abraded during transport and at least partly rounded.

breccia. A rock consisting of cemented, angular, broken rock fragments.

calcite. A widespread, abundant, typically white mineral composed of calcium carbonate ($CaCO_3$); the major component of limestone and marble. Easily scratched with a knife.

caldera. A large, circular or oval basin formed by collapse following a voluminous volcanic eruption.

carbonate rock. A rock composed of the minerals calcite or dolomite, both of which contain carbonate (CO_3); typical examples are limestone, dolomite (dolostone), and marble.

cinder. *See* volcanic cinder.

cinder cone. A cone-shaped accumulation of volcanic cinders erupted from a central basaltic or andesitic vent.

clay. Rock or mineral particles smaller than 0.002 millimeter, or crystals of a clay mineral.

coarse-grained. Said of sedimentary rocks that have particles that are relatively large. Said of igneous rocks with relatively large crystals.

cobble. A rock fragment between 3 and 10 inches in diameter, usually worn to rounded by abrasion in the course of transport.

conglomerate. A sedimentary rock consisting of rounded pebbles, cobbles, or boulders, cemented within a sandy or silty matrix.

contact. The surface between two types or ages of rocks.

crater. A steep-sided, circular depression produced by an explosion.

cross-bedding. Layering in a sedimentary rock that forms at an angle to horizontal, typically the result of currents in water or wind.

crust. The uppermost layer of the Earth's lithosphere, typically 3 to 30 miles thick. Continental crust, which hosts granite, generally consists of rock that is higher in silica and oxygen and lower in iron and magnesium than oceanic crust, which consists mostly of basalt.

debris. A surficial accumulation of loose, broken rock, soil fragments, or both.

debris flow. A relatively fast downslope flowage of mixed rock debris as a wet mass.

deformation. Any process by which preexisting rocks are bent, broken, or uplifted.

desert varnish. A thin coating of dark material abnormally rich in iron and manganese that forms on exposed rock surfaces in desert areas after long exposure.

detachment fault. A low-angle normal fault that allows the rock in its upper plate to deform independent of the rock in its lower plate.

dike. A tabular igneous intrusive body, discordant with the structure of the surrounding rock.

diorite. A group of plutonic rocks intermediate in composition between gabbro and granite, and the approximate intrusive equivalent to the extrusive rock andesite.

dip. The inclination from horizontal of any planar surface within rocks, such as a sedimentary bed, as measured in the steepest direction (e.g., the direction a marble would roll down an inclined surface).

dolomite. A common rock-forming carbonate mineral with the formula $CaMg(CO_3)_2$; term is also applied to a sedimentary rock consisting of dolomite.

erosion. The removal of rock material by natural processes.

extrusive. Said of igneous rock that has erupted onto the surface of the Earth, including lava and pyroclastic material such as volcanic ash.

fanglomerate. The consolidated deposits of an alluvial fan; a variety of conglomerate that is coarse, poorly sorted, only weakly bedded, and contains angular stones.

fault. A fracture along which rock on either side has slipped past each other.

feldspar. A group of common rock-forming minerals composed principally of silicon, aluminum, and oxygen, plus one or more of the elements calcium, sodium, and potassium.

fine-grained. Said of sedimentary rocks that have particles that are relatively small. Said of igneous rocks with relatively small crystals.

fold. Bent or warped rock layers.

foliation. The layering in a metamorphic rock.

formation. Geologically, a rock body of considerable areal extent with consistent characteristics that permit it to be recognized and mapped; not to be confused with outcrop.

fossil. The remains of a plant or animal preserved in a rock.

fossiliferous. Said of a rock containing fossils.

fracture. Any break in rocks caused by natural mechanical failure under stress, including cracks, joints, or faults.

glacier. A large body of natural, land-borne ice that flows.

glaciation. The formation and movement of glaciers or large ice sheets.

gneiss. A regionally metamorphosed rock characterized by alternating bands of coarse mineral grains and finer, flaky mica minerals. Mineral composition is not an essential factor in its definition; varieties are distinguished more by texture and origin.

granite. An intrusive igneous rock consisting mostly of visible crystals of quartz and feldspar.

gravel. An unconsolidated accumulation of rock fragments, mostly of particles larger than sand such as cobbles and pebbles.

groundwater. The water in pores and other openings in subsurface rocks and sediment.

hydrothermal. Pertaining to geothermally heated water.

ice age. A period in Earth history when large sheets of ice covered parts of nonpolar continents.

igneous. Said of rocks or minerals formed by the crystallization of molten material (magma).

intrusive rock. Rock formed by crystallization of magma underground. A body of intrusive rock is said to be an intrusion.

isotope. A species of an element defined by the number of neutrons in its nucleus. Adjective: isotopic.

lakebeds. Fine-grained sedimentary deposits laid down on the floor of a lake.

lava. Extruded magma or the solidified product of such.

limestone. A sedimentary rock composed largely of the mineral calcite $(CaCO_3)$.

magma. Naturally occurring molten rock within the Earth, capable of intrusion and extrusion. Once extruded, it is called **lava**.

marble. A metamorphosed limestone or dolomite, usually coarsely recrystallized.

marine. Pertaining to sedimentary rocks formed in the sea or ocean, and to oceanic environmental conditions.

matrix. Fine-grained rock or mineral particles filling spaces between coarser constituents of a sedimentary rock.

metamorphic rock. A rock that has undergone sufficient mineralogical and physical changes by heat, pressure, or shearing stress to be distinct from the parent rock.

metamorphism. Recrystallization of an existing rock due to heat and pressure within the Earth.

mudstone. A fine-grained sedimentary rock of silt and clay but without the fine-scale layering of shale; hardened mud.

normal fault. An inclined fault on which the block of rock above the fault has moved relatively downward.

obsidian. A black or dark-colored volcanic glass; usually rhyolite lava that cooled too quickly to form crystals.

olivine. An iron and magnesium silicate mineral that typically forms pale-green to pistachio-green crystals in basalt and gabbro.

outcrop. An exposure of bedrock at the Earth's surface. The rock is said to "crop out."

pebble. A rounded small stone, typically waterworn in the course of transport, between 0.17 and 2.5 inches in diameter.

pegmatite. An unusually coarse-grained igneous rock, typically of granitic composition.

plagioclase. A feldspar mineral rich in sodium, calcium, or both. One of the most common rock-forming minerals in igneous and metamorphic rocks.

plate. In a planetary sense, large, drifting plates composing the Earth's solid outer part, the lithosphere. Continental plates are roughly 60 miles thick.

plate tectonics. The movement and deformation of the Earth's crust caused by the interaction of planetary plates.

playa. A smooth, fine-grained lakebed in a desert valley, normally dry but may be flooded.

pluton. An intrusion of igneous rock cooled from magma that rose into Earth's crust.

polygon. A closed geometrical figure with three or more straight sides.

pumice. A silicic, highly bubble-filled volcanic glass, which may be light enough to float on water.

pyroclastic. Said of broken rock material and pumice formed by explosion from a volcanic vent.

quartz. A common rock-forming mineral that is hard, chemically resistant, and composed of silicon and oxygen (SiO_2). Cannot be scratched with a knife.

quartzite. A metamorphic rock formed by the recrystallization of quartz-rich sandstone.

rhyolite. An extrusive igneous rock (lava) of granitic composition, typically light-colored or reddish. Relatively rich in silicon and poor in iron, magnesium, and calcium.

right-lateral fault. A strike-slip fault along which the opposing block moved to the right.

salt pan. A large flat area in which salt water accumulates and evaporates, leaving a layer of snow-white salt crystals.

sandstone. A sedimentary rock composed primarily of rounded or angular, sand-sized particles of rock or mineral, 0.0025 to 0.08 inch in diameter.

scarp. Topographically, a steep slope. Some scarps, called fault scarps, form by recent faulting and may be tens of feet high.

schist. A metamorphic rock characterized by the propensity to break along well-developed layering (foliation), usually involving oriented mica flakes.

sediment. Solid, unconsolidated particulate matter, especially rock detritus that originates by weathering and is transported and deposited by wind or water.

sedimentary rock. Consolidated and typically cemented sediment, characterized by layering called **bedding**.

shale. A finely layered sedimentary rock made of clay, silt, or mud.

shearing. The action and deformation caused by two bodies of rock sliding past each other. The sheared part of a rock is called a **shear zone**.

silt. Fine particulate rock and mineral matter, dust-sized (finer than sand, coarser than clay), between 0.00016 and 0.0025 inch in diameter.

siltstone. A sedimentary rock made primarily of silt.

slip. The amount of horizontal, vertical, or oblique displacement along a fault.

star dune. An individual dune with several ridges radiating out from a central high point, resembling a starfish if viewed from above.

stratigraphy. Pertaining to the stacking or sequence of deposition of sedimentary rock layers and their place in the geologic rock record.

strike-slip fault. A fault along which the relative displacement is horizontal rather than up or down.

syncline. A fold in layered rocks with stratigraphically younger rocks in its core. Most synclines are concave upward (U-shaped).

talc. A hydrated magnesium silicate mineral ($Mg_3Si_4O_{10}(OH)_2$) that is used as a lubricant and thickening agent in myriad industrial applications. In Death Valley, it formed where intruding basaltic magmas reacted with dolomite in the Crystal Spring Formation.

talus. The accumulation of large angular blocks of rock at the base of a cliff.

tectonic. Pertaining to deformation and the forces that cause deformation of the Earth's crust.

tephra. Rock fragments of any size ejected during volcanic eruptions.

thrust fault. A fault with a dip of 45 degrees or less on which the upper block has moved upward relative to the lower block.

tilted fault block. A mountain-scale block of crust that has tilted in the process of moving along a normal fault.

transverse dune. An asymmetric sand dune elongated perpendicular to the prevailing wind direction.

tuff. A volcanic rock formed of consolidated tephra.

turtleback. A comparatively smooth topographic surface with a shape that resembles the shell of a turtle and roughly coincides with a fault zone separating metamorphic rock below from sedimentary and/or volcanic rock above.

vein. A sheet-like deposit of mineral matter within a fracture in rock.

vent (volcanic). A roughly cylindrical opening through which volcanic material is extruded.

volcanic ash. Unconsolidated, explosively fragmented, pyroclastic volcanic material of particle diameter less than 0.079 inch.

volcanic cinder. A porous fragment of lava explosively ejected from a volcanic vent; from pea to baseball size.

weathering. The chemical decomposition and mechanical disintegration of rocks and minerals through interaction with the atmosphere and biosphere.

wind ripple. A wave-like, asymmetric undulation produced in wind-deposited sand.

SOURCES

GENERAL READING

Digonnet, M. 2016. *Hiking Death Valley: A Guide to its Natural Wonders and Mining Past*. Self-published.

Glazner, A. F., A. G. Sylvester, and R. P. Sharp. 2022. *Geology Underfoot in Death Valley and Eastern California*. 2nd ed. Missoula, MT: Mountain Press.

Hunt, C. B. 1975. *Death Valley: Geology, Ecology, Archeology*. Berkeley, CA: University of California Press.

Lingenfelter, R. E. 1986. *Death Valley and the Amargosa: A Land of Illusion*. Berkeley, CA: University of California Press.

Miller, M. B., and L. A. Wright. 2015. *Geology of Death Valley National Park*. 3rd ed. Dubuque: Kendall-Hunt.

TECHNICAL REFERENCES

A BRIEF GEOLOGIC HISTORY OF DEATH VALLEY
(listed here only if not used elsewhere)

Baldridge, W. S. 2004. *Geology of the American Southwest: A Journey through Two Billion Years of Plate-Tectonic History*. Cambridge, UK: Cambridge University Press.

Burchfiel, B. C., and J. H. Stewart. 1966. "Pull-apart" origin of the central segment of Death Valley, California. *GSA Bulletin* 77: 439–432.

Frankel, K. L., A. F. Glazner, E. Kirby, F. C. Monastero, M. D. Strane, and others. 2008. Active tectonics of the eastern California shear zone. In *Field Guide to Plutons, Volcanoes, Faults, Reefs, Dinosaurs, and Possible Glaciation in Selected Areas of Arizona, California, and Nevada*. GSA Field Guide 11, edited by E. M. Duebendorfer and E. I. Smith: 43–81.

Hoffman, P. F. 1991. Did the breakout of Laurentia turn Gondwanaland inside-out? *Science* 252: 1409–12.

Knott, J. R., M. N. Machette, R. E. Klinger, A. M. Sarna-Wojcicki, J. C. Liddicoat, and others. 2008. Reconstructing late Pliocene to middle Pleistocene Death Valley lakes and river systems as a test of pupfish (Cyprinodontidae) dispersal hypotheses. In *Late Cenozoic Drainage History of the Southwestern Great Basin and Lower Colorado River Region: Geologic and Biotic Perspectives*. GSA Special Paper 439, edited by M. C. Reheis, R. Hershlerand, and D. M. Miller: 1–26.

Peng, S. C., L. E. Babcock, and P. Ahlberg. 2020. The Cambrian Period. In *Geologic Time Scale 2020*, edited by F. M. Gradstein, J. G. Ogg, M. D. Schmitz, and G. M.Ogg. Cambridge, UK: Elsevier Publishing: 565–629.

Prave, A. R. 1991. Depositional and sequence stratigraphic framework of the Lower Cambrian Zabriskie Quartzite. In Implications for regional correlations and the Early Cambrian paleogeography of the Death Valley region of California and Nevada. *GSA Bulletin* 104: 505–515.

Rämö, T. O., and J. P. Calzia. 1998. Nd isotopic composition of cratonic rocks in the southern Death Valley region: Evidence for a substantial Archean source component in Mojavia. *Geology* 26: 891–894.

Wright, L. A., B. W. Troxel, E. G. Williams, M. T. Roberts, and P. E. Diehl. 1976. Precambrian sedimentary environments of the Death Valley region, eastern California. In *Geologic Features—Death Valley, California*: California Division of Mines and Geology Special Report 106, edited by B. W. Troxel and L. A. Wright: 7–15.

BLACK MOUNTAINS AND BADWATER BASIN

Blakeley, R. J., R. C. Jachens, J. P. Calzia, and V. E. Langenheim. 1999. Cenozoic basins of the Death Valley extended terrane as reflected in regional-scale gravity anomalies. In *Cenozoic Basins of the Death Valley Region*. GSA Special Paper 333, edited by L. A. Wright and B. W. Troxel: 1–16. Boulder, CO. Site 11.

Brogan, G. E., K. S. Kellogg, D. B. Slemmons, and C. L. Terhune. 1991. *Late Quaternary Faulting along the Death Valley-Furnace Creek Fault System, California and Nevada*. USGS Bulletin 1991. Site 10.

Calzia, J. P., and O. T. Rämö. 2000. Late Cenozoic crustal extension and magmatism, southern Death Valley region, California. In *Great Basin and Sierra Nevada*. GSA Field Guide 2, edited by D. R. Lageson, S. G. Peters, and M. M. Lahren: 135–164. Boulder, CO. Site 13.

Castonguay, S. R. 2013. *Structural Evolution of the Virgin Spring Phase of the Amargosa Chaos, Death Valley, California, USA*. Master's Thesis, University of Oregon. Site 13.

Castonguay, S. R., and M. Miller. 2018. Amargosa Chaos: A product of multiphase deformation. *GSA Abstracts with Programs* 50, Paper No. 2-3. Site 13.

Cuomo, A., and M. Miller. 2023. New findings in the Amargosa Chaos, Death Valley, California, USA. *GSA Abstracts with Programs* 55, Paper No. 253-6. Site 13.

Curry, H. D. 1938. Turtleback fault surfaces in Death Valley, California. *GSA Bulletin* 49: 1875. Sites 7, 11.

de Voogd, B., L. Serpa, L. Brown, E. Hauser, S. Kaufman, and others. 1986. Death Valley bright spot: A midcrustal magma body in the southern Great Basin, California? *Geology* 14: 64–67. Site 12

Evans, J. R., G. C. Taylor, and J. S. Rapp. 1976. *Mines and mineral deposits in Death Valley National Monument, California.* California Division of Mines and Geology Special Report 125. Site 5.

Fleck, R. J. 1970. Age and tectonic significance of volcanic rocks, Death Valley area, California. *GSA Bulletin* 81: 2807–2816. Site 6.

Fleming, Z., T. Pavlis, and G. Trullenque. 2022. Unraveling the multi-phase history of southern Death Valley geology. In *Field Excursions from Las Vegas, Nevada: Guides to the 2022 GSA Cordilleran and Rocky Mountain Joint Section Meeting*. GSA Field Guide 63, edited by G. Jiang and C. Dehler: 67–83. Site 12.

Greene, R. C. 1997. *Geology of the Northern Black Mountains, Death Valley, California.* USGS Open-File Report: 97–79. Site 4.

Hayman, N. W., J. R. Knott, D. S. Cowan, E. Nemser, and A. M. Sarna-Wojcicki. 2003. Quaternary low-angle slip on detachment faults in Death Valley, California. *Geology* 31: 343–346. Site 7.

Hunt, C. B., and D. R. Mabey. 1966. *Stratigraphy and Structure.* USGS Professional Paper 494-A: 162. Death Valley, California. Sites 6, 8, 9.

Hunt, C. B., T. W. Robinson, W. A. Bowles, and A. L. Washburn. 1966. *Hydrologic Basin Death Valley, California.* USGS Professional Paper 494-B. Sites 8, 9.

Knott, J. R., M. N. Machette, E. Wan, R. E. Klinger, J. C. Liddicoat, and others. 2018. Late Neogene–Quaternary tephrochronology, stratigraphy, and paleoclimate of Death Valley, California, USA. *GSA Bulletin* 130: 1231–1255. Site 4.

Knott, J. R., A. M. Sarna-Wojcicki, M. N. Machette, and R. E. Klinger. 2005. Upper Neogene stratigraphy and tectonics of Death Valley: a review. In *Fifty Years of Death Valley Research*, Special issue of Earth Science Reviews, edited by J. P. Calzia: 245–270. Site 11.

Knott, J. R., J. C. Tinsley III, and S. G. Wells. 2002. Are the benches at Mormon Point, Death Valley, California, USA, scarps or strandlines? *Quaternary Research* 58: 352–360. Site 11.

Lowenstein, T. K., J. Li, C. Brown, S. M. Roberts, T. L. Ku, and others. 1999. 200 k.y. paleoclimate record from Death Valley salt core. *Geology* 27: 3–6. Site 9.

Lutz, B. M., J.R. Knott, F. M. Phillips, M. T. Heizler, K. A. Heitkamp, and others. 2022. Tectonically controlled drainage fragmentation in the southwestern Great Basin, USA. *GSA Bulletin* 135: 2296–2314. Sites 1, 2, 3.

McAllister, J. F. 1970. *Geology of the Furnace Creek Borate Area, Death Valley, Inyo County, California.* Map Sheet 14. Sacramento, CA: California Division of Mines and Geology. Sites 1, 2, 3, 5.

Miller, M. 1991. High-angle origin of the currently low-angle Badwater Turtleback fault, Death Valley, California. *Geology* 19: 372–375. Site 7.

Miller, M., and T. L. Pavlis. 2006. The Black Mountains Turtlebacks: Rosetta Stones of Death Valley geology. In *Fifty years of Death Valley Research*, Special issue of Earth Science Reviews, edited by James Calzia: 115–138. Sites 6, 7, 9, 11.

Muessig, S. J., W. M. Pennell, J. R. Knott, and J. P. Calzia. 2019. *Geology of the Monte Blanco borate deposits, Furnace Creek Wash, Death Valley, California.* USGS Open-File Report 2019–1111. Site 3.

Noble, L. F. 1941. Structural features of the Virgin Spring area, Death Valley, California. *GSA Bulletin* 52: 942–1000. Site 13.

Pavlis, T. L., L. F. Serpa, and C. Keener. 1993. Role of seismogenic processes in fault-rock development: An example from Death Valley, California. *Geology* 21: 267–270. Site 11.

Snyder, N. P., and L. L. Kammer. 2008. Dynamic adjustments in channel width in response to a forced diversion, Gower Gulch, Death Valley National Park, California. *Geology* 36: 187–190. Sites 1, 3.

Sohn, M. F., J. R. Knott, and S. A. Mahan. 2014. Paleoseismology of the southern section of the Black Mountains and Southern Death Valley Fault Zones, Death Valley, CA. *Environmental and Engineering Geoscience* 20: 177–198. Site 12.

Troxel, B. W. 1974. Significance of a man-made diversion of Furnace Creek Wash at Zabriskie Point Death Valley, California. In *Guidebook: Death Valley Region, California and Nevada*. Boulder, CO: GSA: 2–16. Site 1.

Wright, L. A., J. Otten, and B. W. Troxel. 1974. Turtleback surfaces of Death Valley viewed as phenomena of extension. *Geology* 2: 53–54. Sites 7, 11.

Wright, L. A., and B. W. Troxel. 1984. *Geology of the Northern Half of the Confidence Hills 15-minute Quadrangle Death Valley Region, Eastern California: the Area of the Amargosa Chaos.* California Division of Mines and Geology Map Sheet 34 and text. Site 13.

NORTHERN DEATH VALLEY

Affinati, S. C., T. D. Hoisch, M. L. Wells, and J. D. Vervoot. 2020. Pressure-temperature-time paths from the Funeral Mountains, California, reveal Jurassic retroarc underthrusting during early Sevier orogenesis. *GSA Bulletin* 132: 1047–1065. Sites 18, 19.

Applegate, J. D. R., and K. V. Hodges. 1995. Mesozoic and Cenozoic extension recorded by metamorphic rocks in the Funeral Mountains, California. *GSA Bulletin* 107: 1063–1076. Site 19.

Bedinger, M. S., and J. R. Harrill. 2012. *Groundwater geology and hydrology of Death Valley National Park, California and Nevada.* Natural Resource Technical Report. NPS/NRSS/WRD/NRTR-2012/652. Fort Collins, CO: National Park Service. Site 14.

Bierman, P. R., and D. R. Montgomery. 2020. *Key Concepts in Geomorphology*. 2nd ed. Austin, TX: MacMillan Press. Sites 16 17.

Hardie, L. A. 1968. The origin of the Recent non-marine evaporite deposit of Saline Valley, Inyo County, California. *Geochimica et Cosmochimica Acta* 32: 1279–1301. Site 30.

Hoisch, T. D., M. L. Wells, M. A. Beyene, S. Styger, and J. D. Vervoort. 2014. Jurassic Barrovian metamorphism in a western US Cordilleran metamorphic core complex, Funeral Mountains, California. *Geology* 42: 399–402. Sites 18, 19.

Hugh, C. R., A. Garcia, J. R. Knott, E. Wan, and M. N. Machette. 2016. *Lake Manly deposits at Three Bare Hills, Park Service Ridge, NPS 5/190 Junction, and Mud Canyon Junction, Death Valley, California*. GSA Cordilleran Section Meeting Abstracts with Programs 48, Paper No. 15-8. Site 15.

Hunt, C. B., and D. R. Mabey. 1966. *Stratigraphy and Structure*. USGS Professional Paper 494-A: 162. Death Valley, California. Site 16.

Hunt, C. B., T. W. Robinson, W. A. Bowles, and A. L. Washburn. 1966. *Hydrologic Basin Death Valley, California*. USGS Professional Paper 494-B. Site 14.

Klinger, R. E. 2001. Beatty Junction Bar Complex. In *Quaternary and Late Pliocene Geology of the Death Valley region: Recent Observations on Tectonics, Stratigraphy, and Lake Cycles*. Pacific Cell, Friends of the Pleistocene trip. USGS Open-File Report 01-51, edited by M. N. Machette, M. L. Johnson, and J.L. Slate: A40–A49. Site 15.

Knott, J. R., J. M. Fantozzi, K. M. Ferguson, S. E. Keller, K. Nadimi, C. A. Rath, and others. 2012. Paleowind velocity and paleocurrents of pluvial Lake Manly, Death Valley, USA. *Quaternary Research* 78: 363–372. Site 15.

Labotka, T. C. 1980. Petrology of a medium-pressure regional metamorphic terrane, Funeral Mountains, California. *American Mineralogist* 65: 670–689. Sites 18, 19.

Mattinson, C. G., J. P. Colgan, J. R. Metcalf, E. L. Miller, and J. L. Wooden. 2007. Late Cretaceous to Paleocene metamorphism and magmatism in the Funeral Mountains metamorphic core complex, Death Valley, California. In *Convergent Margin Terranes and Associated Regions: A Tribute to W.G. Ernst*. GSA Special Paper 419, edited by M. Cloos, W. D. Carlson, M. C. Gilbert, J. G. Liou, and S. S. Sorensen: 205–223. Site 19.

Midttun, N., N. A. Niemi, and B. Gallina. 2023. Stratigraphy of the Eocene-Oligocene Titus Canyon Formation, Death Valley, California, and Eocene extensional tectonism in the Basin and Range. *Geosphere* 23: 258–290. Site 20.

Miller, E. L., M. E. Raftrey, and J. L. Snee. 2022. Downhill from Austin and Ely to Las Vegas: U-Pb detrital zircon suites from the Eocene–Oligocene Titus Canyon Formation and associated strata, Death Valley, California. In *Tectonic Evolution of the Sevier-Laramide Hinterland, Thrust Belt, and Foreland, and Postorogenic Slab Rollback (180–20 Ma)*. GSA Special Paper 555, edited by J. P. Craddock, D. H. Malone, B. Z. Foreman, and A. Konstantinou: 359–378. Site 20.

Owen, L. A., K. L. Frankel, J. R. Knott, S. Reynhout, R. C. Finkel, and others. 2011. Beryllium-10 terrestrial cosmogenic nuclide surface exposure dating of Quaternary landforms in Death Valley. *Geomorphology* 125: 541–557. Site 15.

Reynolds, M. W. 1976. Geology of the Grapevine Mountains, Death Valley, California: A summary. In *Geologic Features: Death Valley Region, California*. California Division of Mines and Geology Special Report 106, edited by B. W. Troxel and L. A. Wright: 19–26. Site 21.

Sauer, K. M. 2014. *Kinematics and timing of superposed deformation in the Funeral Mountains metamorphic core complex*. Master's Thesis. University of Nevada, Las Vegas. Site 19.

Saylor, B. Z. 1991. *The Titus Canyon Formation: Evidence for Early Oligocene Extension in the Death Valley Area, California*. Master's Thesis, Massachusetts Institute of Technology. Site 20.

Wiltschko, D. V., and J. W. Morse. 2001. Crystallization pressure versus "crack seal" as the mechanism for banded veins. *Geology* 29: 79–82. Site 22.

Wright, L. A., and B. W. Troxel. 1993. *Geologic Map of the Central and Northern Funeral Mountains and Adjacent Areas, Death Valley Region, Southern California*. USGS Miscellaneous Investigations Map I-2305. Sites 14, 15, 18.

WESTERN RANGES AND BASINS

Albee, A. L., T. C. Labotka, M. A. Lanphere, and S. D. McDowell. 1981. *Geologic Map of the Telescope Peak Quadrangle*. USGS Geologic Quadrangle GQ-1532. Site 24.

Barba, W. K., M. L. Wells, M. T. Heizler, and E. Turner. 2020. Tectonic evolution of the northern Panamint Range, southeastern California: evidence for synconvergent Late Cretaceous extensional collapse of the Sevier-Laramide orogen. *GSA Abstracts with Programs* 52 (4), Paper No. 27-3. Site 25.

Burchfiel, B. C., K. V. Hodges, and L. H. Royden. 1987. Geology of Panamint Valley-Saline Valley pull-apart system, California: Palinspastic evidence for low-angle geometry of a Neogene range-bounding fault. *Journal of Geophysical Research* 92: 10,422–10,426. Site 30.

Casteel, M. 1986. *Geology of a Portion of the Northwest Last Chance Range, South of Hanging Rock Canyon, Inyo County, California*. Master's Thesis, California State University, Fresno. Site 28.

Champion, D. E., A. Cyr, J. Fierstein, and W. Hildreth. 2018. Monogenetic origin of Ubehebe Crater maar volcano, Death Valley, California:

Paleomagnetic and stratigraphic evidence. *Journal of Volcanology and Geothermal Research* 354: 67–73. Site 26.

Corbett, K., C. T. Wrucke, and C. T. Nelson. 1988. Structure and tectonic history of the Last Chance Thrust System, Inyo Mountains and Last Chance Range, California. In *This Extended Land, Geological Journeys in the Southern Basin and Range*. GSA Guidebook, Cordilleran Section meeting, edited by D. L. Weide and M. L. Faber: 269–292. Site 28.

Crowe, B. M., and R. V. Fisher. 1973. Sedimentary structures in base-surge deposits with special reference to cross-bedding, Ubehebe Craters, Death Valley, California. *GSA Bulletin* 84: 663–682. Site 26.

Fierstein, J., and W. Hildreth. 2017. Eruptive history of the Ubehebe Crater cluster, Death Valley, California. *Journal of Volcanology and Geothermal Research* 335: 128–146. Site 26.

Hall. W. E. 1971. *Geology of the Panamint Butte Quadrangle, Inyo County, California*. USGS Bulletin 1299. Site 31.

Hall, W. E., and E. M. MacKevett Jr. 1962. *Geology and Ore Deposits of the Darwin Quadrangle, Inyo County, California*. USGS Professional Paper 368. Sites 31, 32.

Hodges, K. V., J. D. Walker, and B. P. Wernicke. 1987. Footwall structural evolution of the Tucki Mountain detachment system, Death Valley region, southeastern California. In *Continental Extensional Tectonics*. Geological Society of London Special Publication 28, edited by M. P. Coward, J. F. Dewey, and P. L. Hancock: 393–408. Site 25.

Hunt, C. B., and D. R. Mabey. 1966. *Stratigraphy and Structure*. USGS Professional Paper 494-A: 162. Death Valley, California. Site 23.

Hunt, M. L., and N. M. Vriend. 2010. Booming Sand Dunes. *Annual Review of Earth and Planetary Science* 38: 281–301. Site 29.

Lutz, B. M., J.R. Knott, F. M. Phillips, M. T. Heizler, K. A. Heitkamp, and others. 2022. Tectonically controlled drainage fragmentation in the southwestern Great Basin, USA. *GSA Bulletin* 135: 2296–2314. Site 26.

Jones, R., and R. L. Hooke. 2015. Racetrack Playa: Rocks moved by wind alone. *Aeolian Research* 19: 1–3. Site 27.

Levy, D. A., A. V. Zuza, P. J. Haproff, and M. L. Odlum. 2021. Early Permian tectonic evolution of the Last Chance thrust system: An example of induced subduction initiation along a plate boundary transform. *GSA Bulletin* 133: 1105–1127. Site 28.

Lorenz, R. D., B. K. Jackson, J. W. Barnes, J. Spitale, and J. M. Keller. 2011. Ice rafts not sails: Floating the rocks at Racetrack Playa. *American Journal of Physics* 79: 37–42. Site 27.

Lueders, L. E. 2019. *For the Sake of Salt: A Landscape-Level Management Approach for the Saline Valley Salt Tram*. Master's Thesis, Sonoma State University, California. Site 30.

Mahood, G. A., G. E. Nibler, and A. N. Halliday. 1996. Zoning patterns and petrologic processes in peraluminous magma chambers: Hall Canyon pluton, Panamint Mountains, California. *GSA Bulletin* 108: 437–453. Site 24.

Martin, P., and R.A. Schroeder. 2015. *The Source, Discharge, and Chemical Characteristics of Selected Springs, and the Abundance and Health of Associated Endemic Anuran Species in the Mojave Network Parks*. USGS Scientific Investigations Report 2015-5027. Site 32.

Mase, C. W., S. P. Galanis Jr., and R. J. Munroe. 1979. *Near-surface Heat Flow in Saline Valley, California*. USGS Open-file Report OFR79-1136. Site 30.

McAllister, J. F. 1956. *Geology of the Ubehebe Peak Quadrangle, California*. USGS Geologic Quadrangle Map GQ-95. Site 27.

McDowell, S. D. 1974. Emplacement of the Little Chief Stock, Panamint Range, California. *GSA Bulletin* 85: 1535–1546. Site 24.

McDowell, S. D. 1978. Little Chief Granite porphyry: feldspar crystallization history. *GSA Bulletin* 89: 33–49. Site 24.

Messina, P., and P. Stoffer. 2000. Terrain analysis of the Racetrack Basin and the sliding rocks of Death Valley, California. *Geomorphology* 35: 253–265. Site 27.

Norris, R. D., J. M. Norris, R. D. Lorenz, J. Ray, and B. Jackson. 2014. Sliding rocks on Racetrack Playa, Death Valley National Park: First observation of rocks in motion. *PLOS One 9*, e105948. Site 27.

Oswald, J. A., and S. G. Wesnousky. 2002. Neotectonics and Quaternary geology of the Hunter Mountain fault zone and Saline Valley region, southeastern California. *Geomorphology* 42: 255–278. Site 30.

Pavlik, B. M. 1989. Phytogeography of sand dunes in the Great Basin and Mojave Deserts. *Journal of Biogeography* 16: 227–238. Site 29.

Schweig, E. S. 1989. Basin-Range tectonics in the Darwin Plateau, southwestern Great Basin, California. *GSA Bulletin* 101: 652–662. Site 31.

Stone, P., G. C. Dunne, C. H. Stevens, and R. M. Gulliver. 1989. *Geologic Map of Paleozoic and Mesozoic Rocks in Parts of the Darwin and Adjacent Quadrangles, Inyo County, California*. USGS Miscellaneous Investigations Series Map I-1932. Site 32.

Streitz, R., and M. C. Stinson. 1974. *Geologic Map of California: Death Valley Sheet*. California Division of Mines and Geology. Site 30.

Ver Plack, W. E. 1958. *Salt in California*. California Division of Mines and Geology Bulletin 175. Site 30.

Vriend, N. M., M. L. Hunt, R. W. Clayton, C. E. Brennen, K. S. Brantley, and A. Ruiz-Angulo. 2007. Solving the mystery of booming sand dunes. *Geophysical Research Letters* 34: L16306. Site 29.

Wernicke, B., K. Hodges, and D. Walker. 1986. Geologic evolution of Tucki Mountain and vicinity, central Panamint Range. In *Field Trip Guidebook GSA Cordilleran Section meeting*, edited by G. C. Dunne. Los Angeles, CA: 67–80. Sites 23, 25.

Wrucke, C. T., and K. P. Corbett. 1990. *Geologic Map of the Last Chance Quadrangle, California*. USGS Open-File Report 90–647-A. Site 29.

AMARGOSA VALLEY AND POINTS SOUTH

Bedinger, M. S., and J. R. Harrill. 2012. *Groundwater geology and hydrology of Death Valley National Park, California and Nevada*. Natural Resource Technical Report. NPS/NRSS/WRD/NRTR-2012/652. Fort Collins, CO: National Park Service. Site 40.

Bryne, P. 2012. Pupfish, Downfish: Subterranean tsunami gives vertical shakes to the water-hole home of endangered fishes. *Scientific American*. https://www.scientificamerican.com/article/earthquake-at-devils-hole/ Site 34.

Davidson, C. J. 2021. *Dextral Slip on the Sheephead Gault during Late Miocene Extension, Southeastern California*. Master's Thesis, San Francisco State University. Site 38.

Gibert, L., P. Alfaro, F. J. García-Tortosa, and G. Scott. 2011. Superposed deformed beds produced by single earthquakes (Tecopa Basin, California): Insights into paleoseismology. *Sedimentary Geology* 235: 148–159. Site 36.

Halford, K. J., and T. R. Jackson. 2020. *Groundwater Characterization and Effects of Pumping in the Death Valley Regional Groundwater Flow System, Nevada and California, with Special Reference to Devils Hole*. USGS Professional Paper 1863. Site 40.

Hillhouse, J. W. 1987. *Late Tertiary and Quaternary Geology of the Tecopa Basin, Southeastern California*. USGS Miscellaneous Investigations Map I-1728. Sites 36, 37.

Knott, J. R., M. N. Machette, R. E. Klinger, A. M. Sarna-Wojcicki, J. C. Liddicoat, and others. 2008. Reconstructing late Pliocene to middle Pleistocene Death Valley lakes and river systems as a test of pupfish (Cyprinodontidae) dispersal hypotheses. In *Late Cenozoic Drainage History of the Southwestern Great Basin and Lower Colorado River Region: Geologic and Biotic Perspectives*. GSA Special Paper 439, edited by M. C. Reheis, R. Hershlerand, and D. M. Miller: 1–26. Site 34.

Laczniak, R. J., G. A. DeMeo, S. R. Reiner, J. R. Smith, and W. E. Nylund. 1999. *Estimates of Ground-water Discharge as Determined from Measurements of Evapotranspiration, Ash Meadows Area, Nye County, Nevada*. USGS Water-Resources Investigation Report 99-4079. Site 33.

Lancaster, N., and S. A. Mahan. 2012. Holocene dune formation at Ash Meadows National Wildlife Area, Nevada, USA. *Quaternary Research* 78: 266–274. Site 33.

Larsen, D., and K. Olson. 2019. Evolution of the Pleistocene Lake Tecopa beds, southeastern California: A stratigraphic and sedimentologic perspective. In *From Saline to Freshwater: The Diversity of Western Lakes in Space and Time*. GSA Special Paper 536, edited by S. W. Starratt and M. R. Rosen: 319–357. Site 36.

Lechte, M. A., M. W. Wallace, A. van Smeerdijk Hood, and N. Planavsky. 2018. Cryogenian iron formations in the glaciogenic Kingston Peak Formation, California. *Precambrian Research* 310: 443–462. Site 39.

Le Heron, D. P., M. E. Busfield, and A. R. Prave. 2014. Neoproterozoic ice sheets and olistoliths: multiple glacial cycles in the Kingston Peak Formation, California. *Journal of the Geological Society, London* 171: 525–538. Site 39.

Mahon, R. C., C. M. Dehler, P. K. Link, K. E. Karlstrom, and G. E. Gehrels. 2014. Geochronologic and stratigraphic constraints on the Mesoproterozoic and Neoproterozoic Pahrump Group, Death Valley, California. In A record of the assembly, stability, and breakup of Rodinia. *GSA Bulletin* 126: 652–64. Site 40.

Niemi, N. A., B. P. Wernicke, R. J. Brady, J. B. Saleeby, and G. C. Dunne. 2001. Distribution and provenance of the middle Miocene Eagle Mountain Formation, and implications for regional kinematic analysis of the Basin and Range province. *GSA Bulletin* 113: 419–442. Site 35.

Prave, A. R. 1999. Two diamictites, two cap carbonates, two d^{13}C excursions, two rifts: The Neoproterozoic Kingston Peak Formation, Death Valley, California. *Geology* 27: 339–342. Site 39.

Prave, A. R., and M. R. McMackin. 1999. Depositional framework of mid to late Miocene strata, Dumont Hills and southern margin Kingston Range: Implications for the tectonostratigraphic evolution of the southern Death Valley region. In *Cenozoic Basins of the Southern Death Valley Region*. GSA Special Paper 333, edited by L. A. Wright and B. W. Troxel: 259–275. Boulder, CO. Site 38.

Reheis, M. C., J. Caskey, J. Bright, J. B. Paces, S. Mahan, and E. Wan. 2020. Pleistocene lakes and paleohydrologic environments of the Tecopa basin, California: Constraints on the drainage integration of the Amargosa River. *GSA Bulletin* 132: 1537–1565. Site 36.

Renik, B., N. Christie-Blick, B. W. Troxel, L. A. Wright, and N. A. Niemi. 2008. Re-evaluation of the middle Miocene Eagle Mountain Formation and its significance as a piercing point for the interpretation of extreme extension across the Death Valley region, California, USA. *Journal of Sedimentary Research* 78: 199–219. Site 35.

Reynolds, R. E. 1991. The Shoshone Zoo: A Rancholabrean Assemblage from Tecopa. In *Crossing the Borders: Quaternary Studies in Eastern California and Southwestern Nevada*, edited by R. E. Reynolds. Mojave Desert Quaternary Research Symposium, May 17–20, 1991, San Bernardino County Museum: 158–162. Site 36.

Riggs, A. C., and J. E. Deacon. 2002. Connectivity in desert aquatic ecosystems: the Devils Hole story. In *Spring-fed Wetlands: Important Scientific and Cultural Resources of the Intermountain Region*. DHS Publication 41210: 1–38. Site 34.

Saglam, I., J. Baumsteiger, M. J. Smith, J. L. Casenave, A. L. Nichols, and others. 2016. Phylogenetics support an ancient common origin of two scientific icons: Devils Hole and Devils Hole pupfish. *Molecular Ecology* 25: 3962–3973. Site 34.

Scott, R. K. 1985. *Stratigraphy and Depositional Environments of a Neogene Playa-Lake System, China Ranch Beds, Near Death Valley, California*. Master's Thesis, Pennsylvania State University. Site 38.

Steinkampf, W. C., and W. L. Werrell. 2001. *Ground-water Flow to Death Valley, as Inferred from the Chemistry and Geohydrology of Selected Springs in Death Valley National Park, California and Nevada*. USGS Water-Resources Investigations Report 98-4114. Site 40.

Thompson, R., A. K. Gilmer, K. Souders, and D. P. Miggins. 2022. Miocene magmatism of the central Death Valley Rhombochasm, USA. *GSA Abstracts with Programs* 54 (2), Paper No. 5-1. Site 37.

Tofaif, S., T. M. Vandyk, D. P. LeHeron, and J. Melvin. 2019. Glaciers, flows, and fans: Origins of a Neoproterozoic diamictite in the Saratoga Hills, Death Valley, California. *Sedimentary Geology* 385: 79–95. Site 39.

Troxel, B. W., and E. Heydari. 1982. Basin and Range geology in a roadcut. In *Geology of Selected Areas in the San Bernardino Mountains, Western Mojave Desert, and Southern Great Basin, California*. GSA Cordilleran Section Guidebook, edited by J. D. Cooper, B. W. Troxel, and L. A. Wright: 91–96. Site 37.

Wilson, K. P., M. B. Hausner, and K. C. Brown. 2021. The Devils Hole Pupfish: Science in a Time of Crises. In *Standing between Life and Extinction: Ethics and Ecology of Conserving Aquatic Species in North American Deserts*, edited by David Propst and others. Chicago, IL: University of Chicago Press. Site 34.

Winograd, I. J., T. B. Coplen, J. M. Landwehr, A. C. Riggs, K. R. Ludwig, and others. 1992. Continuous 500,000-year climate record from vein calcite in Devils Hole, Nevada. *Science* 258: 255–260. Site 34.

Winograd, I. J., and F. J. Pearson. 1976. Major Carbon 14 anomaly in a regional carbonate aquifer: Possible evidence for megascale channeling, south central Great Basin. *Water Resources Research* 12: 1125–1143. Site 33.

Woodburne, M. O. and D. P. Whistler. 1991. The Tecopa Lake Beds. In *Crossing the Borders: Quaternary Studies in Eastern California and Southwestern Nevada*, edited by R. E. Reynolds. Mojave Desert Quaternary Research Symposium, May 17–20, 1991, San Bernardino County Museum: 155–157. Site 36.

Wright, L. A., R. C. Greene, I. Çemen, F. C. Johnson, and A. R. Prave. 1999. Tectonostratigraphic development of the Miocene–Pliocene Furnace Creek Basin and related features, Death Valley region, California. In *Cenozoic Basins of the Death Valley Region*. GSA, Special Paper 333, edited by L. A. Wright and B. W. Troxel: 87–114. Site 35.

Mesquite sand dunes.

INDEX

Page numbers in bold face include photographs

archaeological sites, 64, 78
Aguereberry, Pete, 70
Aguereberry Point, 68, **70**–**71**
alkali minerals, 35. *See also* salts
alluvial fans, 4, 17, 23, 25, 27, **29**, 38, 51, 52, 54, 57, **71**, **90**; deposits of former, 12, 25, 30, **32**, 42, 78, 94, 109, **113**, 116; fault scarps on, 3, 35, 39, **36**, **40**, 88; groundwater in, 71, 89; size of, 27, 88
alluvium, 30, 33, 38, 48
alteration, 12, 15, 18, 25, 38, 80, 96
Amargosa Canyon, 99, 110
Amargosa Chaos, 7, 9, 44–45
Amargosa River, 15, 35, 99, 106, 109, 116, 117, 118, 119, 122
Amargose River Canyon, mouth of, 98, 118–20
Amargosa Valley, 12, 18, 22, 98, 99, **107**, 109, 113
Amargosa vole, 111
amphibolite, 56, **57**
aprons, alluvial, 27, 54, 88
Artist Drive Formation, vi, 8, **12**, 13, 17, 18, 20, **22**, 23, **24**, **25**, **26**, 34
Artists Drive, 14, 23–25
Artists Palette, **23**, **25**
ash, volcanic, vi, 12, 16, 18, **29**, 30, 41, 63, 109, 110, **111**. *See also* ash-flow tuff; Bishop Ash; tephra; Yellowstone Ash
ash-flow tuffs, 25, 28, **29**, 112, **113**, **114**
Ashford Mill, 42, 44

Ash Meadows fault zone, 100, 101, **102**
Ash Meadows Fish Conservation Facility, 100
Ash Meadows National Wildlife Refuge, 98, 100–**101**, 102
avalanche deposit, 44

badlands, 12, 16, **17**, **18**, 22, 115
Badlands Junction, 16, 18, 22
Badwater (Basin), 1, 14, 15, 27, 28, **29**, 33, 35–37, **37**, 41, 48, **72**, 74
Badwater Turtleback, **29**, **31**, 32, **33**, 34, 37, 41; fault of, **2**, **31**, **32**; gneisses of, **28**
bajadas, 27, 54, 88
banding: color, **38**, 39; of calcite, **103**; in gneiss, **61**
basalt: lava flows, **12**, 13, 16, 22, 25, **26**, 69, 92, **93**, **94**, **95**, **97**; magma, 78; rim on fragment, **43**; talus, **26**. *See also* cinders
basement, 6, 8, 15, 28, 31, 37, 41, 44, **45**, 110
Basin and Range, 1, 4, **5**, 35, 77, 107
basins, 5, 7, 41, 69, 109, 116, 117. *See also* Badwater Basin; China Ranch Basin; Furnace Creek Basin; Mormon Point Basin; Tecopa Basin
Beatty Bar, 46, 48, **50**–**51**
Beatty Junction, 48, 50, 56
Beck Spring Dolomite, vi, 7, 8, 58, **59**, 116, 122
benches, 40
Billie Mine, 26, 27

biological activity, 10, 54
Bishop, 84
Bishop Ash, 30, 110
Black Mountains, 2, 3, **4**, **5**, **12**, **15**, **17**, **23**, 27, **28**, **31**, **33**, **37**, **43**, **70**; basement rock in, 6; normal faults along, 2, 3–4, 11, 35, 39; southern, 7, 9; talc in, 123. *See also* Black Mountains fault zone; turtlebacks
Black Mountains fault zone, vi, 3–4, 15, 20, **21**, **23**, 25, 27, 30, **31**, **33**, 37, **38**–39, 70
Bonanza King Formation, vi, 8, **10**, 63, 64, **66**, **67**, 70, 82, **105**, 106, 107, 108, 112, 114
borate, 12, **21**, 22, 26, 123; mines, 26
boudins, 58, **59**
Boundary Canyon fault, 2, 47, 59
Butte Valley Formation, vi
brachiopods, 10
breccia, 2, 38, **39**, 62, **66**, **67**

calcite, 66, **67**, **103**, 105
calcium carbonate, 37, 57, 91
calderas, 62, 63, 110, 112
Cambrian, vi, 9, 10, 70
camels, 110
carbonates, 10, 28, 37, 57, 58, 63, 89, 91, 120
Carrara Formation, vi, 8, 64, 70, 84, 85, **106**
caves, 66, 102, 103, 104
celadonite, 12, 18, 25
Cenozoic, vi, 11–12, 69, 76

channel deposits, 116
chemical alteration, 12, 15, 38
Chicago Valley, 113
China Ranch, vi, 98, 99, **115**–**17**
China Ranch Basin, vi, 99, 115, 116, 117
China Ranch Creek, 115, 116, 117
chloride, 28, 89
cinder cones, 4, 42, 43, 92, 94, **95**
Cinder Hill, 42–**43**
cinders, 42, **43**, 92
clay, 12, 16, 18, 37, 50, 52, 86, 116
climate: modern, 1, 48, 105; ancient, 13, 15, 16, 50, 109, 122
colemanite, 26
compression, 11, 47, 64, 69, 92
conglomerate, 25, 51, 62, 63, 109, 117; of Furnace Creek Formation, 12, 13, 16, **17**, **18**, 20, **22**
Conglomerate Mesa, 84
continental plates, 1, 6
continental shelves, 11, 84, 106, 118
Copper Canyon Formation, 8, 39, **40**, 41
Copper Canyon Turtleback, **40**, 41
Cottonwood Mountains, 10, 54, 62, 69, 81, 83, 84, 92
creosote bush, 52
Cretaceous, vi, 11, 92
crinoids, 10
cross-bedding, 64, **65**, 70, **71**, 72
crust (Earth's), 1, 2, 4, 6, 7. *See also* compression; extension
Crystal Pool, 100
Crystal Reservoir, 100, **101**

133

crystals, 33, 35, 57, 58, 60, 61, 66, 82, 92, 97
Crystal Spring Formation, vi, 7, 8, **9**, 44, **45**, 56, **57**, 61, 116, 121, 122, **123**
Curry, Donald, 41
cyanobacteria, 123
Cyprinodon diabolis, 102
Cyprinodon nevadensis nevadensis, 121
Cyprinodon nevadensis shoshone, 111
Cyprinodon salinus, 48

Dantes View, 5, 12, 14, 27–29, **28**, 37
Darwin Canyon Formation, vi, 8, 96, **97**
Darwin Falls, 11, 68, 92, 94, **96**–97
Darwin Plateau, 12, 92
Darwin quartz diorite, 11
date palms, 115
Davidson, Christopher, 117
Daylight Pass, 58
Death Valley: asymmetry of, **43**; climate of, 1; formation of, 12–13; lake in, 13; northern, 47
Death Valley National Park, iii, 1, 69, 72, 99, 102
Death Valley Railroad, 26
debris flows, 46, **54**, **55**, 60, 63, **75**, **76**, 119; fans of, 54–55
deformation, 31, 33, 58, **59**, **60**, 84, 111. *See also* folding
deltas, 9
desert varnish, 54, 55
desert washes. *See* washes, desert
detachment faults, 2, 31, 40, 41, 58. *See also* turtleback faults
Devils Golf Course, 14, **33**, **34**, 35

Devils Hole, 98, 100, 102–5, **103**, **104**, **105**, 111; pupfish, 100, 102
de Voogd, Beatrice, 42
diabase, vi, 7, 8, **9**, 44, 56, 121, 122, **123**
diamictites, 7
dikes, **37**, 92, **94**, **97**
diopside, 97
diorite, **97**
dissolution, 15, 34
diversion channel, **17**
dolomite, 7, 8, 9, 10, 11, 52, 56, **59**, 82, **83**, **123**; in Bonanza King Formation, 66, **67**, 70, 104, 105, 106; in Crystal Springs Formation, 122, 123; metamorphosed, 11. *See also* Beck Spring Dolomite; Noonday Dolomite
dropstones, 118, **119**, **120**
dry falls, 16, 20, **21**, 31, **41**, 58, 60, 61, 76, 96, 97
Dumont Dunes, 87, 119, 121
dune grass, 86
dunes, 4, **9**, **47**, 52–53, 64, 69, 84, **86**–**87**, 88, 91, 98, 100, 119, 121; linear dunes, **86**; star dunes, **52**, 53, 86; transverse dunes, 52, 53, 86, **87**; vegetated, 53, 86. *See also* Eureka Dunes; Kelso Dunes; Mesquite Flat Dunes; Panamint Dunes
dust storms, 18, **47**

Eagle Mountain, 12, 70, **106**–**108**
Eagle Mountain Formation, 22, 106, **107**, 108
earthquakes, 39, 105, 111
Eastern California shear zone, 3
Ely Springs Dolomite, vi, 8
Emigrant Canyon, 71
epidote, 97

erosion, vi, 6, 10, 20, 25, 27, 30, 31, 34, 52, 62, 74, 78, 109; of badlands, 12, 16, 17, 18, 115; by wind, 4, **5**, 52; scouring by, **67**, 75
eruptions. *See* volcanic activity
Eureka Dunes, 52, 68, 69, 86–87
Eureka Mine, 70, 71
Eureka Quartzite, vi, 8, 10
Eureka Valley, 69, 84, 86
evaporates, 27, 28, 89
evaporation, 35, 89, 116, 121
evening primrose, 86
Exclamation Point, 6, 7, 14, 44–**45**
extension, vi, 3, 4, 11, 15, 23, 47, 58, 69, 88, 92, 99, 112, 114. *See also* normal faults; turtleback faults; fault blocks, tilted

facets, triangular, **4**, 38, 88
fanglomerate, 30, 31, **32**, 33, **38**, **115**, **116**
Father Crowley Vista Point, 11, 92–95, 96
faults, 1–4, **2**, **22**, **23**, **31**, **32**, **38**, **39**, **45**, **79**, **85**, **94**, **105**, **108**, **113**; gouge, 31; scarps, 3, 15, 30, **33**, 35, **36**, 38, 39, **40**, 88; striations, **38**; water rising along, 35, 49. *See also* normal faults; *specific fault names*; strike-slip faults; thrust faults; turtleback faults
fault blocks, tilted, 4, 5, 27, **70**, 72, 99, **107**
feldspar, 82. *See also* plagioclase
Fierstein, Judy, 78
fish. *See* pupfish
fissures, 102, 103, 105
flash floods, 75, 89

flooding, 15, 17, **29**, 34, 53, 67, **72**, 86, 121; scouring by, **67**, 75
folding, vi, 2, 16, 34, 47, 48, **60**, 69, 76, **77**
49ers, 13
fossils, 9, 10, 11, 109, 123; mammals, **110**; single-celled organisms, 89
Funeral Formation, vi, 8, 12, **13**, 25, 48, 49, 51
Funeral Mountains, **7**, 10, 11, **12**, **13**, **23**, 47, 57, 58, 62
Furnace Creek Basin, 12, 13, 16, 21, 22, 26, 41, 80
Furnace Creek fault zone, 3, 12, 13, 47, 99
Furnace Creek Formation, vi, 8, 12, **17**, 18, **19**, 20, **21**, **22**, 26, 41, 48, **49**, 57; conglomerate of, 12, **13**, 16, **17**, 18, **22**
Furnace Creek Inn, 12
Furnace Creek Wash, 13, 16, **17**, **23**, 26; alluvial fan of, **71**

garnets, 57, **61**
gas bubbles, 42, 43, 112, 113, **114**
geologic time, vi
glacial deposits, vi, **119**, **120**
glaciation, 1, 9, 118
glauberite, 89
gneiss, 2, **6**, 28, 30, 31, 33, 41, **61**, 110
gold, 11, 13, 56, 57, 60, 69, 70; mines, 56
Golden Canyon, 3, 12, 14, 16, 18–19
Gold Rush, the, 13
gouge, 31
Gower Gulch, 12, 14, 16, 17, 18, 20–22, **21**

Grandstand, the, 82, **83**

Grandview fault, 27

granite, **11**, 61, 72, 74, **82**, 90, 92; cobbles of, **22**, 107, 116, 117; Jurassic, vi, 8, 11, 88, 94, 96; Miocene, vi

Grapevine Canyon, 88, 89

Grapevine Mountains, 10, **47**, 54, 55, 62, 69, 84

gravels: alluvial fan, 4, 27, 30, 50, 51, 52, 71, **113**; granitic, **22**, 107, 116, 117; groundwater in, 48, 89, 96; metamorphic, 33; shoreline, 4, 37, 50, 51. *See also* conglomerates; fanglomerates

Greenwater Range, 12, 13, 26, 107

Greenwater Valley, 4

Greenwater Volcanics, vi, 12, 26, 107

Grenville mountain building, 6

Grotto Canyon, 52; fan, 53

groundwater, 38, 48, 49, 52, 66, 78, 91, 104, 121

gypsum, 28, 89, 115, 116

halite, 28, 89. *See also* salt

Hall Canyon pluton, vi, 11, **74**

Hanging Rock Canyon, 11, 68, 84–**85**

Harrisburg district, 70

Hells Gate, 58

Hidden Valley Dolomite, vi, 8

Hildreth, Wes, 78

hornblende, 97

Horse Thief Springs Formation, vi, 7, 8, **60**, 121, 122, **123**

hot springs, 69, **91**

Hunt, Charles, 28

Hunter Mountain batholith, vi, 8, **11**, 22, 69, 82, **83**, 89, 107; cobbles of, **22**

Hunter Mountain fault, 3, 88, 89

Hurricane Hilary, 29

hydrothermal activity, 6, 11, 22, 25, 69, 96, **97**

Ibex Hills, **121**, **122**, **123**

ice, modern, 81

ice age, 4, 42, 99, 100, 109; icebergs during, 7, 119

igneous rock, 6, 15, 28, 96. *See also* basalt; diabase; granite; pegmatite; rhyolite; volcanic

Independence dike swarm, 97

interbedded, 16, 23, 117

intrusions, 12, 15, 28, 33, 69, 74, 92, 96, 97, 122

invertebrates, 10, 122

Inyo Mountains, 84, 88, 89, **90**

iron oxides, 12, 25, 42, 54, 92, 119

iron stone, 119, 120

Johnnie Formation, vi, 8, 9, 44, **45**, **57**, 58, 72, 74, 116

Jubilee Pass, 44

Jurassic, vi, 11, 69, 97; granitic rock, 88, 89, 92, **94**, **95**, 95. *See also* Hunter Mountain batholith

Keane Wonder fault, 57

Keane Wonder Mine, 7, 11, 47, **56–57**

Keane Wonder Springs, 57

Keeler Canyon Formation, **95**

Kelso Dunes, 87

Kings Pool, **101**

Kingston Peak Formation, vi, 7, 8, 9, 58, **59**, 99, **118**, **119**, **120**

Kingston Range, 116, 117, 118

Kingston Range Granite, 116

Kit Fox Hills, vi

Klare Spring, 63, 64

Knott, Jeff, 40, 50

kyanite, 60

lakebeds, 18, 53, 81, **99**, 109, 110, 112, 113, 114, **117**

lakes: Pleistocene, 4, 13, 50, 99, 109; temporary, 12, 17, 18

laminations, 72, **74**

landslides, 11, 82, 114, 117

Larson, Dan, 109

Last Chance Range, 69, 81, 83, 84, **86**

Last Chance thrust, vi, 11, 84, **85**

lava flows. *See* basalt

lead, 97

Leadfield, 62, 63

left-lateral fault, 2

Lemoigne thrust fault, 11, 94, **95**

Lila C Mine, 26

limestone, 8, 10, 20, 52, 92, **94**, 102, 118; freshwater, 115, 116; metamorphosed, 11, 75; Bonanza King Formation, 106, 112, 114; of Carrara Formation, 70, 106; of Crystal Spring Formation, 7, 56; of Darwin Canyon Formation, 96, **97**

Little Chief stock, vi, **72**, 74

Little Hebe Crater, 78, **79**, 80

Long Valley caldera, 110

Lost Burro Formation, vi, 8

magma, 7, 11, 42, 69, 74, 78, 96, 97, 123

magmatic arc, 92

magnetite, 87

Mahogany Flat Campground, 72

mammals, 110

mammoths, **110**

manganese, 54

Manly, Lake, vi, 4, 13, 37, 42, 50, 51; spit of, **50**, **51**; shorelines of, **37**, **40**

Manly, William, 13

Manly Beacon, **17**, 18

marble, 11, 31, 37, 41, 56, **57**, 58, 60, 75

Maturango Peak, 92

McLean Spring, 48, 49

meander bend, 30, 31, 32

megabreccia, 115, 116

Mesozoic, vi, 11

mesquite, 52

Mesquite Flat, 46, 48, 52, 53, 55; dunes at, 47, **52–53**, 69

metamorphic rock, 6, 15, 31, 32, 33, 40, 41, **45**, 47, 56, 58–61, 68; layering in, 76. *See also* gneiss; marble; migmatite; schist

metamorphism, vi, 6, 9, 28, 40, 47, 57, 82

migmatite, **61**

mining, 63, 97, 110; borate, 26; gold, 56, 69; talc, 7, 122–23

Miocene, vi, 92

mirabilite, 89

Mojave Desert, 6, 86, 100

Mojave National Preserve, 87

Mojave terrane, 6

Monarch Canyon, 7, 11, 46, 47, 58–61, **59**, **60**, **61**

Monarch Canyon fault, 61

Monarch Spring, 58, 61

Mormon Point, 14, 39–41, **40**

Mormon Point Basin, 41

Mormon Point Formation, 30, 41
Mormon Point Turtleback, 28, 39, 41; fault of, **41**
Mormon Point Wash, **40**
Mosaic Canyon, 68, **75–77**, 76; fan of 53
Mosaic Canyon fault, 76, 77
mountain building, vi, 1, 3, 6, 11, 62, 84
mud cracks, **52**, **87**, **101**
mudflats, 109
mudstone, 8, 109, 115, 116
Muessig, Siegfried, 22

Natural Bridge, 2, 30, 31, **32**
Natural Bridge Canyon, 2, 6, 14, 15, 30–33, 40
New York Butte, **90**
Noble, Levi, 44
Noonday Dolomite, vi, 8, 9, 44, **45**, **75**, **76**, **77**, 116, **118**, 120
Nopah Formation, vi, 8, 70
normal faults, 1, **2**, 3–5, 11, 18, 27, 31, 69, 70, 76, 88, 92, 94, 106, 107, **108**, **113**, 118; as magma conduit, 42. *See also* faults, scarps of
Norris, Richard, 81
North American plate, 1
Northern Death Valley fault zone, 3, 13, 55
Northern Death Valley–Furnace Creek fault zone, 3, 13, 46, 47
Nova Formation, vi

obsidian, 112, 113, **114**
Oenothera avita, 86
olivine, 92
orthoclase, **82**

Owens Valley, 13, 89, 110
Owens Valley fault, 3
Owens Valley Formation, 92, **94**
Owlshead Mountains, 11, 14
oxides. *See* iron oxides

Pacific Coast Borax Company, 26, 37
Pacific plate, 1, 3
Padre Point, 92
Pahranagat Mountains, 100
Pahrump Group, vi, 6, **7**, 9, 47, 56, 74, 110
Paleozoic time, vi, 9–10, 11
Palm Springs, 91
Panamint Butte, **95**
Panamint Dunes, 87
Panamint fault, 88
Panamint Range, vi, 4, 5, 6, 9, 10, 11, 27, **29**, **43**, 47, 69, 70, **72**, 74, 92, 94, 123
Panamint Valley, 4, 12, 18, 69, 70, **73**, **74**, 92, 94, 95
Panuga Formation, 62, **63**
Pavlis, Terry, 39
pebbles, 18
pegmatite, 33, 58, **59**
Permian, vi, 11, 84, 92
petroglyphs, 64
pickleweed, 48, **49**
plagioclase, 92, 97
plate tectonics, vi, 1, 6, 13, 34, 84
playas, 11, 15, 16, 18, 21, 22, **69**, 78, **79**, **81**, **82**, **83**, **87**, 88, 89, 91; mud cracks in, **52**, **82**, **87**
Pleistocene, vi, 4, 13, 99, 109, 122
plutons, vi, 11, 39, 74, 89. *See also* granite

Pogonip Group, vi, 8
Point of Rocks Spring, 101
Precambrian, vi, 6–9
Precambrian–Cambrian boundary, 9, 70
precipitates, 11, 27, 35, 66, 71, 91
pumice, 28, 110
pupfish, 13, **48**, 100, 102, 103, **104**, **111**, 121
pyroclastic rock, 92

quarries, 110, 111
quartz, 11, 66, 87; sandstone, 10, 64, 106
quartzite, 8, 9, 10, 11, 56, 60, 64, 65, 66, 70, 71, 84, 85, 106, 108
quartz monzonite, 92, 94, 95

Racetrack Playa, 68, 69, **81–83**
radiometric dating, 103
rain, 15, 16, 28, 34; shadow, 1
Rainbow Canyon, 92, **93**, **94**, **95**
Rainbow Mountain, 115, 116, **117**
red beds, 62, 63, 119
Red Cathedral, 16, 18
Red Pass, 11, 46, 47, **62**–63
Resting Spring Pass, 98, 110, **112**–14
Resting Spring Range, **99**, 107, **109**, 110, 112, 114
Rest Spring Shale, vi, 8, 84, 85
reverse faults, 2
rhyolite lava, 25
rifting, 7, 9, 118
right-lateral faults, 1, 2, 42, 56, 88, 116
ripple marks, 18, **19**, 72
ripples, **52**, **87**

rivers: deposits of, 7, 9, 11, 17, 18, 22, 62, 64, 107, 111. *See also* Amargosa River
rock art, 64
rock fall, 66
Rodinia, vi, 6, 7, 9, 47, 118
Rogers Peak, **72**, 74
Ryan, 12, 14, **26**, 27

Saline Valley, 68, **69**, **88**–91
Saline Valley Salt Company, 89
salt, 4, 15, 27, 28, 34, **35**, 36, 37, **40**, **69**, 71; plants tolerant of, 48, 49. *See also* salt pan
Salt Creek, 35, 48, 49; pupfish, 48
Salt Creek Hills, 46, 48–**49**
Salt Creek Spring, 48, 49
Salt Lake, **88**, 89, 91
salt pan, 27, 28, 34, 35, **36**, 37, **69**, 70, 71, **72**
San Andreas fault, 1, 3, 42
sand, 15, 18, 35, 37, 41, 52, 71, 86, 87, 91, 111, 118; abrasion by, 4, **5**. *See also* dunes; sandstone
sand dunes. *See* dunes
sandstone, 7, 8, 56, 62, 63, 72, 77, 96, 109, 115, 118, 122; ash-rich, 63; basalt-rimmed, 42, **43**; fine-grained, 12, 16, 18, **21**, **22**; metamorphosed, 11; quartz, 10, 64, 106; ripplemarks in, **19**. *See also* cross-bedding; Zabriskie Quartzite
Saratoga Spring, 7, **121**–23
scarps, fault, 3, 15, 30, **33**, 35, **36**, 38, 39, **40**, 88
schist, 6, 11, 56, **57**, 58, 59, 60
sedimentary rock. *See* conglomerate; dolomite; fanglomerate; limestone; sandstone; siltstone

seiches, 105
seismic waves, 105
shale, vi, 7, 8, 10, 11, 56, 84, 85, 106, 118, 119
shearing, vi, 3, 33, 58, 60
Sheephead fault zone, 98, 99, 116, 117
shoreline deposits, 4, 15, **37**, **40**, 50
Shoshone (people), 1
Shoshone (Village), 98, 99, 106, **109**–11
Shoshone Museum, **110**
Shoshone pupfish, **111**
Shoshone Spring, 111
Shoshone Volcanics, 12
Sierra Nevada, 1, **73**, 97
silica, 91, 97
sills, 7, 56
silt, 4, 15, 35, 50, 71, 75, 111, 119
siltstone, 7, **17**, **19**, 20, **21**, 62, **63**, 70, 72, 77, 106, **115**, 118, 119; badlands in, 12, 16, 18, 22; permeability of, 48, 49
silver, 97
Skidoo pluton, 11
Skidoo mining district, 11, 69, 71
Smith Mountain, 12
Smith Mountain Granite, vi, 11, 39
Snowball Earth, 118, 120
Southern Death Valley fault zone, 3, 42, **43**
Sperry Hills, 115, 119
spits, 4, 50, 51
springs, 35, 39, 48, 49, 61, 71, 89, 96, 100, 101, 103, 110, 111, 116, **121**; deposits of, 109; hot, 69, 91. See also specific spring names
Spring Mountains, **5**, 121
stamp mills, **56**, 57
staurolite, 57
Stirling Quartzite, vi, 8, 9, 58, 72, 77
storms, 4, 15, 17, 47, 53, 54, 99
Stovepipe Wells, 18, 46, 53, 68, 76
stratigraphic column, 8
streams. See Amargosa River; Furnace Creek Wash; rivers; Salt Creek
striations, 23, 119
strike-slip faults, 1, **2**, 23, 44, 88, 99
stromatolites, **123**
subduction, 1, 11, 92
sulfates, 28, 89
supercontinents, 6, 7, 47, 118
Swallenia alexandrae, 86
Swansea, 89

talc, vi, 7, 8, **9**, 121, 122, **123**
talus, 26, 62, 66
Tecopa, 87, 98, 109, 110, 111, 115
Tecopa, Lake, 109; beds of, **109**, 110
Tecopa Basin, vi, 109
tectonic activity, vi, 1, 6, 13, 34, 84
Telescope Peak, 1, 3, 27, 68, **72**–**74**, 92
tephra, 78, 80
thrust faults, vi, **2**, 11, 47, 69, 84, **85**, 94, 95
till, 7, **119**
Timber Mountain caldera, 62, 63
Timbisha Shoshone, 1
Tin Mountain Limestone, vi, 8
titanotheres, 11
Titus Canyon, 10, 46, 47, 64–**67**
Titus Canyon anticline, **64**–65
Titus Canyon Formation, vi, 8, 11, 47, **62**, **63**
tramway, 56, 57, 89
transverse dunes, 52, 53, 86, 87
Transverse Ranges, 97
travertine, 28, 57, 91
triangular facets, **4**, 38, 88
trilobites, 10
Troxel, Bennie, 44, 110
Tucki Mountain, 47, 52, 53, 71
tufa, 37
tuffs. See ash-flow tuffs
turtlebacks, 28, **31**, **33**, 34, 37, 39, **40**, 41; faults of, vi, 2, 30, **31**, **32**, 39, **40**, **41**

Ubehebe Crater, 4, 42, 69, 78–**80**
Ubehebe Peak, **82**, 83
ulexite, 26
uplifts, vi, 4, 6, 11, 25, 28, 47, 94
Upper Warm Spring, 89, **91**

veins, 11, 66, **103**
ventifacts, 4, **5**
vents, 92
volcanic activity, 12, 22, 23, 42–43, 78–81, 92, 99
volcanic ash. See ash, volcanic
volcanic bombs, 80
volcanic glass, 112
volcanic rock, 8, **12**, 15, **31**, 62, 63, 107; pebbles of, 18, 20, 32, 33. See also Artist Drive Formation; ash-flow tuffs; basalt; cinders; Furnace Creek Formation
Vriend, Nathalie, 87
Wahguyhe Formation, 62, **63**
washes, desert, 13, 15, 16, 17, 23, 26, 40, 41, 96
weathering, 74
welded ash, 28, **114**
wetlands, 100, 121
Willow Spring, 115
Willow Spring pluton, vi, 11, 39
wind, 4, 5, 18, 34, 47, 50, 52, 53, 81, 86, 87, 91, 109
wineglass canyons, **4**, 15, 20, **23**, **33**, 47, **90**
Wood Canyon Formation, vi, 8, 9, 64, **65**, **70**
Wright, Lauren, 44

Yavapai mountain building, vi, 6
Yellowstone ash, **111**

Zabriskie Point, 12, 14, 16–**17**, 18, 22
Zabriskie Quartzite, vi, 8, 10, 64, **65**, 66, **70**, **71**, 84, **85**, 106, 108
zinc, 97

—MORGAN JUDY PHOTO

Marli Miller spent most of her career studying the geology of Death Valley, beginning with her times as a volunteer in the park's mining office (now Resource Management) in 1984 and 1986. She completed her BA in geology at Colorado College in 1982 and her MS and PhD in structural geology at the University of Washington in 1987 and 1992, respectively. She is a renowned photographer and a senior instructor and researcher at the University of Oregon, where she teaches a variety of courses, including introductory geology, structural geology, field geology, and geophotography. In addition to numerous technical papers, she is the author of *Roadside Geology of Oregon, Roadside Geology of Washington* with coauthor Darrel Cowan, and *Geology of Death Valley National Park* with coauthor Lauren A. Wright. Marli has two daughters, Lindsay and Megan.